RAND NATIONAL DEFENSE RESEARCH INSTITUTE

T0170207

# Assessing and Evaluating Department of Defense Efforts to Inform, Influence, and Persuade

## Worked Example

Christopher Paul

Prepared for the Joint Information Operations Warfare Center

For more information on this publication, visit www.rand.org/t/RR809z4

Library of Congress Cataloging-in-Publication Data
ISBN: 978-0-8330-9733-0

Published by the RAND Corporation, Santa Monica, Calif.

© Copyright 2017 RAND Corporation

RAND® is a registered trademark.

Cover: Photo by 1st Lt. Henry Chan.

Support RAND
Make a tax-deductible charitable contribution at
www.rand.org/giving/contribute

www.rand.org

# Preface

This short report provides a worked example of the approach to the assessment and evaluation of U.S. Department of Defense (DoD) efforts to inform, influence, and persuade as detailed in *Assessing and Evaluating Department of Defense Efforts to Inform, Influence, and Persuade: Desk Reference* and *Assessing and Evaluating Department of Defense Efforts to Inform, Influence, and Persuade: Handbook for Practitioners.*[1] Both volumes were developed as part of the project "Laying the Foundation for the Assessment of Inform, Influence, and Persuade Efforts," which sought to identify and recommend selected best practices in assessment and evaluation drawn from existing practice in DoD, academic evaluation research, public relations, public diplomacy, and public communication, including social marketing. As a companion to the handbook and desk reference, this report provides a concrete practical example of the planning and application of assessment and evaluation practices and principles in a realistic but fictional scenario context.

This worked example should be useful to practitioners charged with planning, executing, and assessing DoD efforts to inform, influence, and persuade.

This research was sponsored by the Joint Information Operations Warfare Center and conducted within the International Security and Defense Policy Center of the RAND National Defense Research Institute, a federally funded research and development center sponsored by the Office of the Secretary of Defense, the Joint Staff, the Unified Combatant Commands, the Navy, the Marine Corps, the defense agencies, and the defense Intelligence Community.

For more information on the International Security and Defense Policy Center, see www.rand.org/nsrd/ndri/centers/isdp.html or contact the director (contact information is provided on the web page).

---

[1] Christopher Paul, Jessica Yeats, Colin P. Clarke, and Miriam Matthews, *Assessing and Evaluating Department of Defense Efforts to Inform, Influence, and Persuade: Desk Reference*, Santa Monica, Calif.: RAND Corporation, RR-809/1-OSD, 2014; Christopher Paul, Jessica Yeats, Colin P. Clarke, Miriam Matthews, and Lauren Skrabala, *Assessing and Evaluating Department of Defense Efforts to Inform, Influence, and Persuade: Handbook for Practitioners*, Santa Monica, Calif.: RAND Corporation, RR-809/2-OSD, 2014.

# Contents

Preface ........................................................................................................... iii

Figures and Tables ........................................................................................ vii

Summary ........................................................................................................ ix

Acknowledgments .......................................................................................... xi

Abbreviations .............................................................................................. xiii

CHAPTER ONE

**Introduction and Purpose** ......................................................................... 1

Structure and Layout of This Report ............................................................ 3

CHAPTER TWO

**Review of Relevant Assessment Principles** ............................................... 5

Effective Assessment Starts in the Planning Phase ........................................ 5

Above All Else, Assessment Must Support Decisionmaking ........................... 6

Effective Assessment Requires Clear, Realistic, and Measurable Goals ........... 6

Effective Assessment Requires a Theory of Change or Logic of the Effort Connecting
    Activities to Objectives ............................................................................ 9

Be Thoughtful About What You Measure ....................................................12

CHAPTER THREE

**Scenario: MNF-DP's Takanwei Campaign** ...............................................17

Background: Takanwei and the DLB ...........................................................17

Relevant Elements of the MNF-DP Campaign ........................................... 23

CHAPTER FOUR

**Worked Example of Assessment Design and Planning for Selected Campaign
    Elements** ..............................................................................................35

Effective Assessment Requires Clear, Realistic, and Measurable Goals ..........35

Effective Assessment Requires a Theory of Change/Logic of the Effort Connecting
    Activities to Objectives ............................................................................ 40

Be Thoughtful About What You Measure ....................................................50

CHAPTER FIVE
**Conclusions and Review**................................................................59

APPENDIXES
A.  **Telecommunications and Media in Takanwei**...........................61
B.  **Timeline and Road to Crisis, January–July 2022** .......................63

**References**................................................................................67

# Figures and Tables

## Figures

2.1. Logic Model Template . . . . . . . . . . . . . . . . . . . . . . . . . . . . . . . . . . . . . . . . . . . . . . . . . . . . . . . . . . 11
3.1. Map of the Dunarian Peninsula and Surrounding Region . . . . . . . . . . . . . . . . . . . . 18

## Tables

3.1. Crosswalk of Subordinate Objectives and Planned IRC Tasks . . . . . . . . . . . . . . . . . 27
4.1. First Step in Building a Logic Model for Objective 1 . . . . . . . . . . . . . . . . . . . . . . . . . 43
4.2. Second Step in Building a Logic Model for Objective 1 . . . . . . . . . . . . . . . . . . . . . . 44
4.3. Third (and Final) Step in Building a Logic Model for Objective 1, with
a Key Change Highlighted . . . . . . . . . . . . . . . . . . . . . . . . . . . . . . . . . . . . . . . . . . . . . . . . 46

# Summary

The U.S. Department of Defense (DoD) spends more than $250 million per year on information operations (IO) and information-related capabilities for influence efforts at the strategic and operational levels. How effective are those efforts? Are they well executed? How well do they support military objectives? Are they efficient (cost-effective)? Are some efforts better than others in terms of execution, effectiveness, or efficiency? Unfortunately, generating assessments of efforts to inform, influence, and persuade (IIP) has proved to be challenging across the government, including DoD. Challenges include difficulties associated with observing changes in behavior and attitudes, lengthy timelines to achieve impact, causal ambiguity, and struggles to present results in ways that are useful to stakeholders and decisionmakers.

Previous RAND research, published in handbook and desk reference formats, distilled and synthesized insights and advice for improving the assessment of DoD IIP efforts and programs, drawing on a comprehensive literature review and more than 100 interviews with subject-matter experts.[2] This report expands on those previous publications by providing a worked example: an extended, concrete discussion of planning for IIP assessment in the context of a realistic military operation. It complements those earlier reports; ideally, the reader has read the handbook (or will read it in parallel with this report) and will refer to the desk reference for detailed explanations of the principles, concepts, and terms illustrated here.

To increase the accessibility of this work and provide a more immersive experience for the reader, the example is presented as input to the assessment planning activities of a fictitious narrator, MAJ John Fnorky, the J39 IO officer on an equally fictitious joint task force conducting the fictitious Operation Gathered Resolve in the fictitious country of Takanwei. This structure permits the use of realistic operational and planning details without the distraction of an actual historical or potential future operation

---

[2] Christopher Paul, Jessica Yeats, Colin P. Clarke, and Miriam Matthews, *Assessing and Evaluating Department of Defense Efforts to Inform, Influence, and Persuade: Desk Reference*, Santa Monica, Calif.: RAND Corporation, RR-809/1-OSD, 2014; Christopher Paul, Jessica Yeats, Colin P. Clarke, Miriam Matthews, and Lauren Skrabala, *Assessing and Evaluating Department of Defense Efforts to Inform, Influence, and Persuade: Handbook for Practitioners*, Santa Monica, Calif.: RAND Corporation, RR-809/2-OSD, 2014.

or context—avoiding debates about the veracity of accounts of events and threats to the sensitivity of operational details.

Readers follow MAJ Fnorky as he leads his IO working group through three stages of the assessment planning process. They begin by reviewing and refining their initial objectives to ensure that these objectives are SMART: **s**pecific, **m**easurable, **a**chievable, **r**elevant, and **t**ime-bound. They then build (and refine) a logic model that captures the logic of the effort connecting plans for their IIP efforts with the objectives refined in the first stage. Finally, they develop strategies for data collection and assessment to measure the progress of their IIP efforts.

Readers (and practitioners) are reminded of several core principles of assessment. The principles are listed and discussed, but they are also concretely illustrated as part of the worked example. These principles include the fact that **effective assessment starts in the planning phase**, and that, above all else, **assessment must support decision-making**. The discussion emphasizes that **effective assessment requires clear, realistic, and measurable goals**—goals that are SMART, specify the observable behaviors sought, and articulate a target threshold for measuring change or progress. These goals or objectives need to at least imply what failure would look like, and they should be able to be broken into smaller subordinate objectives or sequential steps to make assessment easier. Evaluating progress against these objectives requires some kind of baseline measurement. Further, **effective assessment requires a theory of change or logic of the effort connecting activities to objectives** and including planned inputs, activities, outputs, and outcomes. Readers are advised that logic models can help identify possible constraints, barriers, or unintended consequences to planned activities and that logic models can either start small and grow or start big and be pruned. Good target audience analysis can help avoid bad assumptions in logic models, and a "fail fast" implementation approach can help identify flawed assumptions and provide guidance to correct them. Finally, **assessors should be thoughtful about what they measure**. Logic models can provide a framework for selecting and prioritizing measures. When choosing measures, assessors should consider the relative importance of different candidates and be sure to collect indicators of both success and failure. Measures should not conflate exposure and effectiveness where messaging is concerned, and they should aim to capture trends over time. Assessors should use multiple data sources to triangulate information where possible but should conserve scarce resources by measuring only as precisely as required.

# Acknowledgments

This report owes a great debt to the many experts and colleagues whose good ideas and insightful observations were captured by this broader research effort. I am particularly grateful for the advice and support of personnel at the Joint Information Operations Warfare Center, the project sponsor: Rick Grimes, Charlie Chenoweth, Jay Harris, Ron Thornton, and Patrick Grady. Particular and special thanks to Charlie Chenoweth for suggesting both the name and notion of the fictional narrator of the report, MAJ Fnorky. I also must thank the sponsors of the foundational research on which this effort builds, COL Dan Ermer and Paula Trimble, both formerly of the Rapid Reaction Technology Office; LTC Albert Armonda, formerly in the Office of the Under Secretary of Defense for Policy (OUSD[P]), IO; and Austin Branch, former director of OUSD(P) IO. During RAND's quality assurance process, I substantially improved this report based on the comments and input of the two reviewers, Scott Savitz and Todd Helmus. The manuscript was further improved by a vigorous edit from Lauren Skrabala (with the support of production editor Beth Bernstein). Finally, this report (indeed, this whole stream of research) would not have been possible without my RAND colleagues and coauthors of the previous foundational research in this area: Jessica Yeats, Colin Clarke, Miriam Matthews, and Lauren Skrabala.

# Abbreviations

| | |
|---|---|
| C2 | command and control |
| CA | civil affairs |
| CMO | civil-military operations |
| COA | course of action |
| DLB | Dunarian Liberation Brotherhood |
| DoD | U.S. Department of Defense |
| EW | electronic warfare |
| GOKA | Government of the Kingdom of Arpanda |
| GOT | Government of Takanwei |
| IED | improvised explosive device |
| IIP | inform, influence, and persuade |
| IO | information operations |
| IOWG | information operations working group |
| IRC | information-related capability |
| JOPP | joint operation planning process |
| JP | joint publication |
| JTF | joint task force |
| MILDEC | military deception |
| MISO | military information support operations |
| MNF-DP | Multi-National Forces–Dunarian Peninsula |

| | |
|---|---|
| NGO | nongovernmental organization |
| PA | public affairs |
| PSF | Takanwei Public Safety Force |
| SMART | specific, measurable, achievable, relevant, and time-bound |
| TAA | target audience analysis |
| TCO | transnational criminal organization |
| UN | United Nations |
| VEO | violent extremist organization |

# CHAPTER ONE

# Introduction and Purpose

Achieving key national security objectives demands that the U.S. government and U.S. Department of Defense (DoD) effectively and credibly communicate with and influence a broad range of foreign audiences. For this reason, it is important to measure the performance and effectiveness of inform, influence, and persuade (IIP) efforts that support larger military campaigns. Thorough and accurate assessments of these efforts guide their refinement, ensure that finite resources are allocated efficiently, and inform accurate reporting of progress toward DoD's goals. Such efforts represent a significant investment for the U.S. government: DoD spends more than $250 million per year on information operations (IO) and information-related capabilities (IRCs) to support IIP efforts at the strategic and operational levels. How effective are those efforts? Are they well executed? How well do they support military objectives? Are they efficient (cost-effective)? Are some efforts better than others in terms of execution, effectiveness, or efficiency? The answers to these questions are not clear.

Unfortunately, generating assessments of such activities has been a challenge across the government and DoD. IIP efforts often target the cognitive dimension of the information environment, attempting to effect changes in attitudes and opinions. These changes can be quite difficult to observe or measure accurately. Even when activities seek to influence *behavior* (more easily observable and thus more measurable), causal conflation remains a challenge. Previous RAND research sought to support DoD progress in this area. Drawing on a comprehensive literature review and more than 100 interviews with subject-matter experts, the project "Laying the Foundation for the Assessment of Inform, Influence, and Persuade Efforts" distilled and synthesized insights and advice for improving the assessment of DoD IIP efforts and programs. These results were published in handbook and desk reference formats.[1] This report expands on those previous reports by providing a worked example: an extended,

---

[1]   Christopher Paul, Jessica Yeats, Colin P. Clarke, and Miriam Matthews, *Assessing and Evaluating Department of Defense Efforts to Inform, Influence, and Persuade: Desk Reference*, Santa Monica, Calif.: RAND Corporation, RR-809/2-OSD, 2014; Christopher Paul, Jessica Yeats, Colin P. Clarke, Miriam Matthews, and Lauren Skrabala, *Assessing and Evaluating Department of Defense Efforts to Inform, Influence, and Persuade: Handbook for Practitioners*, Santa Monica, Calif.: RAND Corporation, RR-809/2-OSD, 2014.

concrete discussion of planning for IIP assessment in the context of a realistic military operation. This report is intended as a companion to those earlier reports, and, ideally, the reader will have read the handbook (or will read it in parallel) and will have the desk reference available for additional background while reading this report.

The previous reports distilled and compiled best practices for the assessment and evaluation of IIP efforts across several sectors: defense, government more broadly, industry, and academia. Inputs to the research included more than 100 subject-matter interviews across sectors and the review of hundreds of reports, articles, textbooks, white papers, and examples. The findings and recommendations fall into 11 categories:

- motivation for assessment and evaluation
- assessment best practices and principles
- challenges to organizing for assessment
- determining what is worth measuring
- developing measures
- assessment design and implementation
- formative evaluation
- surveys and sampling
- measurement and data collection
- presenting and using assessment
- developing an organizational culture of assessment.

This report presents a realistic worked example demonstrating the application of a prioritized selection of best principles and practices, summarized in Chapter Two.

The goal is to illustrate in a concrete and practical manner how the identified principles might be implemented in actual practice. The worked example builds on a fictional operation and an artificial scenario context. The operation and scenario context are an amalgam of several existing joint and service exercise scenarios (to minimize the extent to which the author had to imagine absolutely everything about the operation), but the countries, locations, and operations are not drawn from any real-world place or events, nor from any exercise scenarios currently or previously in use. A wholly fictional mission in a wholly fictional environment avoids possible sensitivities associated with discussing real historical or contemporary adversaries or potential future adversaries.

To further increase the accessibility of presentation and to facilitate engagement with the primary audience (those with IIP assessment roles at a major staff or headquarters), the material is presented in the voice of a fictitious narrator—one who is more like the intended audience than is the author. MAJ John Fnorky often addresses readers in the first person (from his own point of view) or in second person, speaking to the reader as a colleague concerned with or engaged in these types of activities.[2] MAJ Fnorky's interjections are set in italics and help guide the reader in interpreting and applying the principles, practices, and lessons he shares throughout the report. In the next passage, he makes his own introduction:

*Hi. I'm MAJ John Fnorky, U.S. Army. I've been asked to serve as the J39, the IO officer, in Joint Task Force–Operation Gathered Resolve, as the lead headquarters in support of Multi-National Forces–Dunarian Peninsula. I've known about the assignment and the operation for less than 72 hours. Since I found out, I've been scrambling to download and pull off the shelves everything I can that might be useful to me while I'm downrange. I've been learning a whole bunch about Takanwei, Arpanda, and some of the regional partners we'll be working with. One of my buddies at the Joint Information Operations Warfare Center sent me a couple of reports on IO assessment, so those are in my stack, too. As an experienced FA30 I feel pretty well spun-up on IO and the IRCs, but assessment is one of those things we haven't always nailed shut in the past, so I'll give that a quick look and see if we can do better this time out. I'll keep you posted.*

## Structure and Layout of This Report

The remainder of this report proceeds as follows. Chapter Two reviews the assessment principles identified in the previous research that are demonstrated and applied in later chapters. Chapter Three describes the fictional mission and scenario context: the Multi-National Forces–Dunarian Peninsula (MNF-DP) campaign in and near the notional countries of Takanwei and Arpanda, including the planned IIP efforts that will need to be assessed. Chapter Four presents the worked example of assessment planning, demonstrating the processes for preparing and implementing assessments of these operations and discussing some of the results. Chapter Five reviews the results and offers concluding comments and lessons for future—real-world—application.

---

[2]  The author's previous experience with instructional fiction suggests that such an approach can be quite effective. U.S. military readers are encouraged to seek out Christopher Paul and William Marcellino, *Dominating Duffer's Domain: Lessons for the 21st-Century U.S. Marine Corps Information Operations Practitioner*, Santa Monica, Calif.: RAND Corporation, RR-1166-1-OSD, 2016, or Christopher Paul and William Marcellino, *Dominating Duffer's Domain: Lessons for the 21st-Century U.S. Army Information Operations Practitioner*, Santa Monica, Calif.: RAND Corporation, RR-1166/1-A, 2017.

# CHAPTER TWO

# Review of Relevant Assessment Principles

*Fnorky here again. So, I read a bunch on how to do assessment and evaluation, and I cannot wait to try to apply those principles in actual practice. Keeping in mind what I know about the campaign we're planning against the DLB (that's the Dunarian Liberation Brotherhood), I've cribbed what I think are the best bits from the assessment readings for us to try using in our operations in Takanwei and Arpanda. Listen up and I'll share.*

This chapter summarizes key findings and lessons from *Assessing and Evaluating Department of Defense Efforts to Inform, Influence, and Persuade: Desk Reference* and *Handbook for Practitioners* that are relevant and applicable to planning and designing assessments of IIP efforts for MNF-DP in the artificial scenario context used here. These principles are demonstrated concretely and in context in the scenario introduction in Chapter Three and the worked example presented in Chapter Four.

## Effective Assessment Starts in the Planning Phase

Assessment does not just happen: You have to plan for it to happen. Assessment needs to start in the planning phase for two reasons. First, you have to ensure that the goals and objectives for your operations and activities are specified correctly during planning so that they can be assessed. (Too often, the goals that are set are too abstract and cannot be meaningfully measured.) Second, assessment activities must be planned alongside other operational activities so that needed data collection becomes part of the plan and actually happens.

According to the joint operation planning process (JOPP) framework laid out in Joint Publication (JP) 5-0, assessment should be considered at the earliest stages.[1] Formative assessment may inform operational design during mission analysis. Preliminary assessment plans should be included in course-of-action (COA) development and

---

[1]   U.S. Joint Chiefs of Staff, *Joint Operation Planning*, Joint Publication 5-0, Washington, D.C., August 11, 2011.

should be wargamed along with other COA elements during the COA analysis and wargaming step.

## Above All Else, Assessment Must Support Decisionmaking

Assessment can serve a number of different purposes, including supporting planning, improving processes, and making resource decisions. One thing is common to all assessment purposes: Assessment must support decisionmaking. Assessment divorced from decisionmaking has no value. If formative assessment prior to planning cannot bring about changes to plans, do not bother. If process assessment identifies ways to be more efficient or effective but those new approaches cannot be adopted, then the assessment was a waste. If resources are going to be allocated based on a political process that does not consider the results of performance assessment, then the assessment does not matter.

In the context of this report—planning assessment for a joint task force (JTF)—the purpose of the assessment is to guide process improvement and increase effectiveness. Any assessment to support *planning* will have already been completed; assessment of accountability and costs may be necessary but will likely take a backseat to operational effectiveness. Key decisions about IIP activities will need to be made, including whether to alter or adjust activities, whether to expand or build upon activities, whether to proceed to the next phase in a sequence of activities, or whether to terminate a set of activities. Assessment should be planned with the decisions it needs to support in mind. Ideally, assessment plans (like other plans) will have a timeline with specific decision points called out for commanders and staffs.

## Effective Assessment Requires Clear, Realistic, and Measurable Goals

One of the reasons assessment must start in planning is to ensure that the planning process produces goals and objectives that can be measured and assessed. Effective assessment requires clear, realistic, and measurable goals. If the initial objectives provided are too vague to assess against, try to define them more precisely and then push them back to superiors for discussion and confirmation. In JOPP, most of the elements of operational design should take place as part of step 2, mission analysis.[2] During mission analysis is when objectives should be articulated and refined, in concert with higher headquarters, if necessary. Clear objectives should be an input to mission analy-

---

[2]  JOPP formally has seven steps: (1) planning initiation, (2) mission analysis, (3) COA development, (4) COA analysis and wargaming, (5) COA comparison, (6) COA approval, and (7) plan or order development. For further detail, see JP 5-0 (U.S. Joint Chiefs of Staff, 2011).

sis, but if the objectives are not clear, mission analysis should provide an opportunity to seek refinement.

As described in JP 5-0, operational art is about describing the military end state that must be achieved (ends), the sequence of actions that are likely to lead to those objectives (ways), and the resources required (means). Operational design is the part of operational art that combines an understanding of the current state of affairs, the military problem, and the desired end state to develop the operational approach. These are the four steps in operational design:

1. understand the strategic direction
2. understand the operational environment
3. define the problem
4. use the results of steps 1–3 to develop a solution—i.e., the operational approach.

The third chapter of JP 5-0, "Operational Art and Operational Design," urges commanders to collaborate with their higher headquarters to resolve differences in the interpretation of objectives to achieve clarity. This should occur as part of the "understand the strategic direction" element of operational design, and it should be part of the first (planning initiation) or second (mission analysis) step of JOPP—or perhaps between them. This exhortation applies not only to the commander and higher head-quarters but also to the assessment planner and the planning team lead.

Setting objectives for an IIP effort or activity is a nontrivial matter. While it is easy to identify high-level goals that at least point in the right direction (e.g., "win," "stabilize the province," "promote democracy"), getting from ambiguous aspirations or end states to useful objectives is challenging. Clear objectives are necessary for both the design and execution of effective IIP efforts *and* for their assessment. The following sections describe some of the challenges and tensions inherent in setting IIP objectives and offer some advice for considering and setting objectives.

## Objectives Should Be SMART

That is, objectives should be **s**pecific, **m**easurable, **a**chievable, **r**elevant, and **t**ime-bound. It is important that objectives specify *what* is to be accomplished, not *how* it is to be accomplished. As noted in JP 5-0, "An objective does not infer ways and/or means—it is not written as a task."[3] Similarly, objectives need to state (or clearly imply) *when* the change needs to take place and to *what extent* (how many people, how much change).

---

[3]   U.S. Joint Chiefs of Staff, 2011, p. III-20.

### Good IIP Objectives Should Specify the Observable Behaviors Sought and from Whom They Are Sought

IIP objectives need to specify what behavior or behavior change is desired and from what audience or group. *Whom* do you want to *do what*? Clarity and precision regarding audiences and behaviors make it much easier to observe and measure whether the desired effects are being realized, and they make it easier to think about the extent to which intermediate objectives will actually contribute to higher-level objectives.

### Evaluating Change Requires a Baseline

When setting an objective that requires change and specifying how much change is required, there is an implicit need to understand conditions in the information environment before IIP activities begin (the baseline conditions). If you want to be able to calculate the delta, or change, in a condition or behavior, you need to have measured a baseline against which to compare. It does little good to launch an effort to get 60 percent of eligible residents to participate in town meetings if 70 percent of residents already do so. Similarly, it would be impossible to evaluate the effectiveness of a gun buyback program without some baseline information about how many weapons are held by residents in an area.

### Target Thresholds: How Much Is Enough?

Part of being specific about an objective is setting some sort of target. No realistic effort aspires to get everyone in a group to completely embrace a desired behavior. No matter what, not everyone is going to vote, not everyone is going to call a tip line, and not everyone is going to comply with a curfew. The target threshold should be set based on what is required in order to meet operational objectives. How much is enough? That threshold can then be presented as a percentage, a ratio, or a fixed amount.

### Good Objectives Need to at Least Imply What Failure Would Look Like

When thinking about how much of a behavior or behavior change is sufficient to support operational objectives, you should also think about what an insufficient amount would be. If the target threshold is the minimum amount to positively contribute to success, is anything less than that failure? Is there a gray area where achievement is positive but not sufficiently positive to support broader operational goals? Is the lower bound of that gray area a point at which you can be confident that the outcome is *not* contributing positively to broader operational goals? Could limited, sub-threshold change even be counterproductive?

### Break Objectives into Smaller Subordinate Objectives or Sequential Steps

Even if you know you need a certain level of accomplishment to succeed, you may not need that all at once, or achieving it may not be realistic in the short term. If you break up your objective into phases and then set incremental goals, you can still track progress. If expected progress lags, you can make course corrections before reaching the end of your planned effort and discovering that you have failed. Incremental objectives also provide an opportunity to manage expectations about the pace of change. For example, you may expect little or no change during the first few weeks of an effort as you prepare and distribute influence products. Then, as the mass of products available increases and awareness and exposure go up, change may (slowly) start to occur. Finally, you may expect a landslide of behavior change as early adopters are seen to receive whatever positive benefit was offered and many others in the target audience begin to conform.

Breaking objectives into "bite-sized," incremental subordinate objectives can make it easier to articulate a logic model or a theory of change (as discussed in the next section).

## Effective Assessment Requires a Theory of Change or Logic of the Effort Connecting Activities to Objectives

As mentioned earlier, operational art is about describing the military end state that must be achieved (ends), the sequence of actions that are likely to lead to those objectives (ways), and the resources required (means). This specification of ends, ways, and means sounds very much like an articulation of a theory of change (or a logic model).[4]

Articulated at the outset, during planning, a theory of change/logic of the effort can help clarify goals, explicitly connect planned activities to those goals, and support the assessment process. A good theory of change will also capture possible unintended consequences or provide indicators of failure—helping you identify where links in the logical chain have been broken by faulty assumptions, inadequate execution, or factors outside your control (disruptors).

IIP activities require more specification of the logic of the effort than do most kinetic activities. When using artillery or air strikes to destroy a bridge, for example, it is not necessary to specify the logic of the effort: Everyone has a good intuition of the fact that explosions and impact weaken structures, and, ultimately, it is just physics. You may need an engineer or weaponeer to help you understand exactly how much of what types of fires will be required to collapse a specific bridge, or to choose specific aim points, but it is clear what kind of expert you need to consult. This is much less clear for IIP activities. Psychology and cognition are not governed by well-structured laws, like those of physics. There are competing theories about influencing both individuals and populations, as well as a range of experiences and cultural, linguistic, societal, contextual, or other factors that can determine how efforts to influence human

---

[4] Logic models are depictions of how an effort or initiative is supposed to work and are described in greater detail in the next section.

dynamics play out. When working to influence, inform, or persuade, it is imperative that you spell out the logic of the effort and explicitly state how you believe your activities will lead to the intended results. As part of the theory of change or logic of the effort, you will specify your assumptions, identify vulnerable assumptions, and plan to use assessment to validate those assumptions (or prove them wrong so that you can make different assumptions, alter the effort, and move forward).

**Logic Model Basics: Inputs, Activities, Outputs, and Outcomes**

One of the classic ways to build and then depict a theory of change is in a logic model. Logic models traditionally list inputs, activities, outputs, and outcomes. See Figure 2.1 for a sample template of a logic model.[5] The *inputs* to a program or effort are the resources required to conduct the program. Of course, inputs include personnel and funding, but they are usually articulated in more specific terms—perhaps the specific expertise required or the number of personnel (or person-hours of effort) available. An effort's *activities* are the verbs associated with the use of the resources, and they are the undertakings of the program. Activities might include planning, design, and dissemination of messages or products. They could also include any of the actions necessary to transform inputs into outputs. *Outputs* are produced by conducting the activities with the inputs. Outputs include traditional measures of performance and indicators that the activities have been executed as planned. These might include execution and dissemination indicators, measures of reach, measures of receipt/reception, indicators of participation, and so on. *Outcomes* (or effects) are the state of a target population that the effort is expected to have changed.

This is the result of the process: The inputs resource the activities, and the activities produce the outputs. The outputs lead to the outcomes. This last step is a critical juncture from a theory of change perspective, as the mechanism by which the outputs (messages disseminated, messages received) connect to the outcomes (behaviors changed) is critical and a potentially vulnerable assumption in influence and persuasion. Outcomes are characteristics or behaviors of the audience or population, not of the program or effort. The outputs are related to the program or effort, and they describe the products, services, or messages it provides. Outcomes refer to the results (or lack of results) of the outputs produced, not just their delivery or receipt. Optionally, logic models can also list disruptors—the things you have identified that might interfere with the progression from inputs to activities to outputs to outcomes.

---

[5] Logic models are discussed in greater detail in the companion desk reference and handbook (in Chapter Five, in both cases).

**Figure 2.1**
**Logic Model Template**

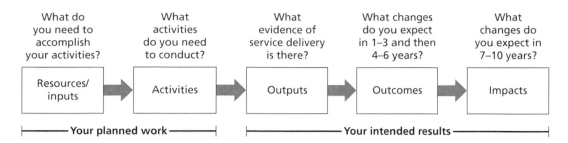

SOURCE: Donna M. Mertens and Amy T. Wilson, *Program Evaluation Theory and Practice: A Comprehensive Guide*, New York: Guilford Press, 2012, p. 245, Figure 7.1. Used with permission.
RAND *RR809/4-2.1*

JP 5-0 both explicitly and implicitly follows logic models. For each of the elements of operational design and each of the JOPP steps, JP 5-0 explicitly lists the inputs to that element or step and the expected outputs. In both processes, many of the outputs of earlier steps or elements become inputs to later steps. The overall presentation supports a logic model framework. For example, the emphasis in operational art on ends, ways, and means corresponds to logic model language: The ends are the outputs and outcomes, the ways are the activities, and the means are the inputs.

### Building Logic Models: Start Big and Prune, or Start Small and Grow

There is at least as much art as science to achieving the right level of detail in a logic model or theory of change. To a certain degree, the framework of *inputs to activities to outputs to outcomes to impacts* is sufficient to begin developing a logic model. Begin at the right, with SMART objectives, and work backward to the left. What has to happen for those objectives to be met? What do you need to do to make those things happen? What resources do you need to do those things?

Logic modeling is almost always an iterative process. Initial models are often either too big, with more detail than needed for planning purposes, or too small, with insufficient detail. However, it is important to make a first attempt, while acknowledging that there will be a need to either expand or prune it (or possibly a little of both in different sections) until it is good enough. Trying to plan activities and assessments based on the draft logic model will help identify elements in need of attention in subsequent iterations.

### Identify and Watch for Possible Constraints, Barriers, Disruptors, and Unintended Consequences

Traditionally, logic models do not list constraints, barriers, or disruptors that can impede the program being described. However, these traditional program logic models rarely accounted for complex operating environments or adaptive adversaries. While building an IIP logic model (and perhaps as part of the model), it is critical to identify and collect information on things that could go wrong. These things could include vulnerable assumptions that might prove to be incorrect, supply chain or other capability shortfalls that might constrain inputs, or things that could happen in the environment, either due to chance or due to enemy action, that could interfere with the intended logic of the effort. By identifying these constraints or vulnerabilities you can both (1) make contingency plans to overcome or surmount barriers and (2) plan monitoring so you get early indicators that barriers or disruptors are materializing.

### Target Audience Analysis Can Help You Avoid Bad Assumptions

Effective target audience analysis (TAA) is essential to effective IIP efforts because it helps identify the right lines of persuasion for a specific group. The basics of the process are laid out in doctrine and are not belabored here. In addition to helping identify lines of persuasion likely to be effective for an audience, TAA can also help avoid bad assumptions. Comparing a logic model with the results of TAA or getting feedback on a logic model from a relevant cultural subject-matter expert can help you find flaws in the logic model before operations commence.

### "Fail Fast" as a Solution to Bad Assumptions in Logic Models

Even after operations commence, you can and will find flaws in a logic model. The only imperative under these circumstances is to "fail fast." If there are vulnerable assumptions or uncertainties in the logic model informing your efforts, try to operate under those assumptions, closely monitor and assess results, and plan to make quick corrections if uncertainties resolve unfavorably. When time and resources allow, pilot testing an effort on a limited scale can be a low-cost form of "fail fast."

## Be Thoughtful About What You Measure

One of the places where assessment planning can break down, even when good assessment principles are understood and applied, is in the selection of measures. No matter how well you have designed your assessment framework, if you measure the wrong things, you will not be able to use the resulting information to make good decisions. Often, circumstances will force you to accept a proxy measure or an indicator for the underlying concept you want to observe. Suppose you are interested in how much support there is among a particular population segment for a terrorist or insurgent group. You might be able to survey members of that group about their attitudes, but it might be impossible to conduct such a survey (nonpermissive security environment), or you might suspect that respondents will not answer that question (supporting the insurgents is illegal, and supporters fear arrest or reprisal because of their responses). If you are unable to measure these attitudes directly, one or more proxy measures or indicators might suit. For example, you might look at the number of social media posts from the population of interest that express sentiments favorable to the insurgents. Alternatively or in combination, you could consult intelligence reports on the insurgents' freedom of movement (under the assumption that villages the insurgents visit must at least tolerate, if not actively support, their presence).

If the chosen indicators do not actually represent the underlying construct, assessment can fail. Because of these risks, it is tempting to try to measure many, many things, reasoning that some of the measurements will pan out. Although it is based on sound reasoning, this approach does not consider the costs (in time, manpower, and money) associated with measurement. When "metric bloat" prevails, it can complicate analysis or make data collection unacceptably burdensome. Be thoughtful about what you plan to measure, trying to identify what you really need and considering the difficulty and costs associated with measuring it.

## Logic Models Provide a Framework for Selecting and Prioritizing Measures

A logic model encapsulates a theory of change/logic of the effort and, done well, suggests things to measure. One might ask,

- Were all of the resources needed for the effort available? (inputs)
- Were all activities conducted as planned? On schedule? (activities)
- Did the activities produce what was intended? Did those products reach the desired audience? What proportion of that audience? (outputs)
- What proportion of the target audience engaged in the desired behavior? With what frequency? (outcomes)
- What things prevented the activities from leading to the outputs, or the outputs from leading to the outcomes? (barriers or disruptors)

These questions point directly to possible measures and also help you prioritize. A good logic model provides a good initial set of candidate measures.

## Consider the Importance of Candidate Measures

Another factor to consider is how useful or valuable a measure will be. In his book *How to Measure Anything*, Douglas Hubbard cleverly argues that the value of information is a function of two factors: the uncertainty associated with the information (how confident you are that the information is correct) and the cost (not just monetary costs, but personnel costs and other consequences) of the information being wrong.[6] For example, the cost of being wrong would be high if an influence campaign hinged on the

---

[6]  Douglas W. Hubbard, *How to Measure Anything: Finding the Value of "Intangibles" in Business*, Hoboken, N.J.: John Wiley and Sons, 2010.

assumption that changing villagers' attitudes toward insurgents will lead to increased reporting of insurgent activity to a tip line. Because the costs would be high if that assumption were wrong, measures that demonstrate the connection between attitude change and tip line use—and thus reduce uncertainty—would be very valuable. When identifying things to measure, give priority to "loadbearing" and vulnerable cause-and-effect relationships in the logic model. The things you are least certain about and the things that are most vulnerable to enemy action or other contextual factors are the things that you most need to measure. Prioritize those things.

## Avoid the Temptation to Collect Data Only on Indicators of Success

Remember that assessment is at its best when it helps you diagnose and repair a struggling effort. Therefore, rather than only monitoring and measuring positive progress as expected according to the logic model, you must be alert to (and monitor and measure) possible obstacles along the way. Consider the list of barriers or disruptors you identified. What is the threshold or target you identified for success? Expand on that. What would failure look like? How will you know if it is coming? Measure those things.

## Do Not Conflate Exposure to a Message with Effectiveness

In IIP, whether an audience receives the message is a key part of the logical chain that leads to behavior change. For someone to receive a message, it must be available, via whatever media it is conveyed. However, just because a message is present in the information environment where an individual resides, you should not assume that the individual has received it, and even if someone has received a message, that does not mean he or she has been persuaded by it. Neither *reach* nor *exposure* is an adequate proxy for *effectiveness*, though both might be worth measuring.

## Triangulate from Multiple Data Sources Where Possible

The best evaluations use many measures and different methods and data sources to obtain more reliable results. This is referred to as *triangulation*, where multiple intersecting measurements provide greater confidence than a single measure.[7] The most valid observations are those that converge across multiple qualitative and quantitative measurements. Triangulation is particularly important for things that are difficult to measure directly or difficult to measure with confidence. Suppose you are assessing a military information support operations (MISO) effort and want to know whether an audience is receiving radio broadcasts related to the effort. If you do not have the capability to survey the audience, you might measure a number of different things instead.

---

[7] The term has its origin in surveying, using a single known length and measuring two angles to determine the third angle and the remaining side lengths of a triangle. The term is also used (and follows the same general logic) in radio direction finding. Alone, a single sensor detecting a signal can indicate the direction of the signal; working in tandem with a second sensor in a second location, operators are able to triangulate a single broadcast location from two direction-only measurements.

You could conduct spot checks to confirm reception quality throughout the audience's area; monitor the audience's social media use for mentions of the broadcasts, or their themes, and messages; review key leader engagement after-action reports, civil affairs (CA) after-action reports, or patrol after-action reports for mention of the broadcasts or broadcast themes; or search for mentions of the broadcasts or themes in various intelligence sources or reports. Using just one of these proxy measures might be insufficient (or inconsistent), but using several (triangulating) would give you greater confidence in results. In other words, when the reliability of individual indicators is low, the combined reliability of multiple indicators of the same measure will be better.

### Try to Capture Trends Over Time

The most valid and useful measurements are those that capture trends over time and across areas. At minimum, evaluating change requires a baseline and then a result measurement. Multiple measurements over time plotted against time produce a trend line, which can show incremental progress (or lack of progress) toward a goal. By measuring incremental progress over greater periods of time, even more can be learned, such as seasonal variations, the lag between your efforts and their effects, or the impact of adversary efforts or exogenous shocks.

### Only Measure as Precisely as Required

Once you start identifying possible measures, it can be hard to stop. Having too many measures will stress the resources of your data collection structure—whether your sources are intelligence reports, observations for executing forces, contracted data collection initiatives, other sources, or a combination of several sources. One way to control the burden on collection is to think about how precisely you need to measure things. In some cases, you want precise information, such as the number of products or messages delivered or broadcast. However, in other cases, you do not need as much precision and a rough idea is good enough. Many things that are inputs are in this category, as are many of the barriers or disruptors. For example, if the adversary is jamming your broadcasts, you do not need to know the exact percentage of broadcasts getting through, but you would benefit from some sort of stoplight assessment of both the reach and quality of the broadcasts. Just like many inputs can be measured at the stoplight or go/no-go level, many possible disruptors can be measured sufficiently as problem/not a problem. Remember that assessment ultimately supports decisionmaking. How much precision do you really need about a construct to make relevant decisions?

*So, those are the high points from my reading on assessment. I'm going to keep all that in mind as I convene the IO working group as part of the JTF's planning process for operations in and around Takanwei. I want to make sure we get assessment considerations rolled in right from the start and that we plan to include as much best practice as we can.*

# CHAPTER THREE

# Scenario: MNF-DP's Takanwei Campaign

*Here's the situation that led to our JTF and the assembly of MNF-DP: Beginning in January 2022, the DLB, a violent extremist organization (VEO) residing primarily in Takanwei, carried out a series of direct and indirect operations against traffic in the Straits of Arpanda, causing significant regional instability and turmoil to global shipping markets. The DLB remains a significant threat to the security of the Straits of Arpanda— a threat beyond the control of the Government of the Kingdom of Arpanda (GOKA) and the Government of Takanwei (GOT). On June 30, 2022, at the request of the GOKA and the GOT, the United Nations (UN) Security Council passed Security Council Resolution 16090, authorizing under Chapter VII a multinational force, led by the United States, to assist Arpanda and Takanwei in neutralizing the DLB and restoring secure and stable conditions at a level that each host nation can sustain.*

This chapter provides the relevant background on the artificial scenario that serves as the foundation for the worked example that follows in Chapter Four. This background begins with a description of Takanwei and the threat posed by the DLB. It concludes with a discussion of the elements of the planned MNF-DP campaign against the DLB under the purview of MAJ Fnorky, illustrating assessment planning and design for an IIP effort.

## Background: Takanwei and the DLB

The primary area of responsibility for this operation encompasses the two countries that share the Dunarian Peninsula—the Kingdom of Arpanda and the Republic of Takanwei—as well as the Straits of Arpanda, an essential sea line of communication for vessels seeking to transit the adjacent Dunarian Sea. Figure 3.1 shows a map of the region. This section profiles the notional country of Takanwei and the notional VEO that poses a threat to its security, the DLB.

**Figure 3.1**
**Map of the Dunarian Peninsula and Surrounding Region**

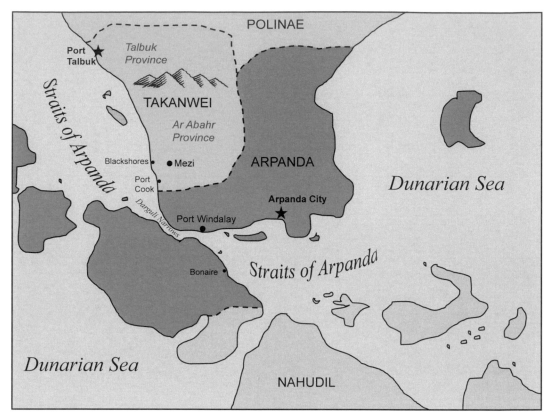

## Takanwei

Takanwei sits on the Dunarian Peninsula, north and west of Arpanda and north of the narrowest portions of the Straits of Arpanda. Takanwei's capital is Port Talbuk, in the northwest of the country. The country's second-largest city is Mezi, located in the southwest. As described below, Takanwei has a rich culture and history.

### Geography

Takanwei is a lightly populated country with around 380,000 people and two distinct geographic regions. The center consists of a *jungle mountain range, which effectively divides the north from the south,* posing a challenge for overland travel.

### Government

Takanwei is a *struggling democratic state without a viable security establishment that has friendly relations with Arpanda and other regional countries.* Its security forces are undermanned, undersupplied, and poorly led. There are high rates of lawlessness and violence in the country.

Since the end of the civil war in 2002, Takanwei has held two successful presidential elections, in 2008 and 2016. The 2022 presidential election has been canceled as a direct result of the DLB's disruptive activities in Takanwei's southern Ar Abahr Province. The last round of congressional elections took place in 2018 and left a legislature hopelessly divided among three competing parties.

The country's political process has been corrupted by partisans vying for advantages over each other. Additionally, the DLB wields an unusual degree of influence in southern Takanwei, in Ar Abahr Province, due to the group's success in bribing and intimidating local political, judicial, and law enforcement officials.

### History

Years of *civil war in the 1990s and early 2000s left Takanwei in a state of ruin and nearly unmanageable.* The armed conflict pitted guerrillas against national security forces, aided by militias.

### Population

Takanwei's diverse population of around 380,000 people is generally neutral in its interactions with U.S. citizens and the citizens of other nations. In the past, this response has grown more positive with offers of financial or infrastructure assistance. Local and municipal leaders largely favor and encourage visits and assistance from foreigners, especially investors and nongovernmental organizations (NGOs). The population is customarily respectful to government officials, security forces, and NGO personnel.

The exception to the population's neutral-to-positive interactions with foreigners comes sporadically from about a third of the country's rural regions and the poorest urban communities. Residents of these may misunderstand the intentions of foreigners and react less favorably to the presence of large groups of U.S. and other foreign nationals in their neighborhoods, particularly security forces assisting the Takanwei government. Violence is not expected in these areas, but demonstrations and other expressions of ill will are probable.

Further, the DLB is working to alter the outlook of the population in its favor and to promote distrust of the government and foreigners. With the DLB's growing influence, the attitudes of certain populations in southern Takanwei have begun to shift to a position of pro-DLB and anti–U.S./foreign nationals. These unfriendly populations can be found in the six southern communities of Ar Abahr Province where the DLB has established semipermanent military bases. These communities have also turned against the government, the Takanwei Public Safety Force (PSF), and unusual aircraft, but not NGOs (yet).

Information about telecommunications and news media in Takanwei can be found in Appendix A of this report.

**Dunarian Liberation Brotherhood**

The DLB is widely regarded as a VEO. The group supported and fought alongside the guerrillas in Takanwei's civil war in the 1980s and 1990s. Small numbers of DLB fighters have remained active in the country since the war ended and the primary opposition guerrillas disarmed and demobilized. Now, the DLB is actively threatening the Straits of Arpanda and disrupting the flow of regional commerce, attempting to destabilize countries throughout the region, and expanding its current holdings in Takanwei. The DLB's stated objectives include a desire to reduce the corrupting influence of Western capitalism by disrupting the flow of commerce through the Straits of Arpanda while maintaining and expanding its territory in Takanwei.

The DLB has established training and logistics bases in Takanwei and operational centers in Arpanda and elsewhere in the region. From these operational centers, the DLB generates ground threats in Takanwei and Arpanda and interdicts merchant shipping in or en route to the Straits of Arpanda using semi-submersibles, Q-ships (armed merchant ships with concealed weaponry), and waterborne improvised explosive devices (IEDs). DLB presence and activity are concentrated in three areas: coastal Arpanda, southern Takanwei, and selected islands near and along the Straits of Arpanda.

Intelligence sources have reported on the group's composition and recent activities:

- The DLB has reconstituted its forces in several areas and established a stronghold in Takanwei's southern Ar Abahr Province, where it seeks to destabilize the government and establish a power base from which it can attack the Straits of Arpanda.
- While seeking to attack the straits, the DLB is engaged in other offensive operations in the region, including ground assaults, assassinations, bombings, disruptive attacks at sea, infrastructure building, smuggling, propaganda, recruitment, and training.
- The DLB's current strength stands at roughly 3,000 personnel; around 2,400 are located in military camps in Takanwei, at least 300 are distributed throughout Arpanda, and the remainder are located on various islands near the straits.
- Small DLB commando units, numbering 25–30 personnel each, appear to be operating in Arpanda. Covertly dispersed among the Arpandian population, these units carry out attacks and disruptive operations against Arpanda and the straits.
- To raise funds, the DLB engages in criminal activities, including smuggling illicit cargo.
- The DLB conducts operations with other criminal organizations in the region to leverage larger smuggling networks and outside expertise. Currently, in Takanwei, the DLB is cooperating with the Termina Triad and the Opal Organization.
- DLB propaganda consists of anti-government and anti-capitalist rhetoric, recruiting information, and speeches from its leaders. It draws largely on popular com-

plaints against government and foreigners interspersed with manufactured facts and figures.

DLB is also involved in several other activities that confirm its status as a VEO:

- criminal activities
- smuggling
- extortion
- kidnapping
- money laundering
- counterfeiting
- human trafficking
- robberies and theft
- recruitment
- insurgency training
- IED manufacture and use
- cooperation with other VEOs
- IO
- cyber operations
- ground assaults
- assassinations
- intimidation strikes
- acts of piracy.

### DLB Operations in and Through the Information Environment

The DLB employs a number of IRC-equivalents, and its most robust capability area is propaganda.

The group actively seeks outside expertise to acquire additional asymmetric capabilities in such areas as cyber and electronic warfare. It encourages the usual types of cyberattacks by third parties, such as distributed denial-of-service attacks on government websites. However, there is no indication that the DLB has an organic offensive cyber capability. A known third-party actor who frequently advocates cyberattacks on behalf of the DLB goes by the handle Red Spyder, but little else is known about this actor. In addition, information technology students at Mezi University most likely have some basic hacking and phishing capabilities, but their link to the DLB has not been clearly established. However, the DLB has employed several university students to manage its web presence.

The DLB deploys commercially available GPS and VHF radio jammers to both mask its own navigational systems and disrupt legitimate communication channels.

### DLB Propaganda

The DLB makes regular use of propaganda and employs a wide range of platforms, including social media, radio broadcasts, face-to-face interactions, and print products. The primary mode of dissemination is the Internet—through its official website, leadership Twitter feeds, YouTube videos, known hacker forums, email accounts, and the group's Facebook page. It also relies heavily on social media and encrypted communication apps to coordinate attacks and monitor its members. It is believed that most of the group's activities are planned and directed from Mezi University in Takanwei.

In addition to its online activities, the DLB engages in face-to-face marketing campaigns involving old-fashioned pamphlets, comic books, posters, urban graffiti, couriers, and direct radio broadcasting. The group is known to possess at least three mobile radio broadcast units. Since most governments employ radio direction-finding equipment, the DLB frequently moves its broadcast equipment among various remote locations in Takanwei. Mezi University is also known to air pro-DLB public-service radio broadcasts on its own radio station. The Takanweian government estimates that the vast majority of DLB recruits originate in the pro-DLB belt of the country's south. DLB propaganda products are often found in Mezi as well. In the north, DLB propaganda material is extremely common in Talbuk and nearby communities.

Beyond these traditional propaganda modes, the DLB engages in what could be labeled "propaganda of the deed." In smaller towns and villages, the group actively organizes community social events to bolster its position. It has also detonated explosive devices on small boats in the straits in demonstrations accompanied by warnings that the continued presence of international forces will lead to "damage" to international shipping. The DLB has employed several platoons of loyalists to start small fires and detonate IEDs in urban areas as a show of resolve.

In terms of content, DLB propaganda typically consists of anti-government and anti-capitalist rhetoric, recruiting information, and speeches from the group's leaders. These strongly worded propaganda pieces draw largely on popular complaints against the current government and foreigners. However, the propaganda is deeply interspersed with illegitimate "facts," figures, and political arguments. The DLB employs various propaganda campaigns to legitimize its cause and promote recruitment. The Takanweian government indicates that the DLB's online recruiting efforts have been most successful in the larger urban areas of the south. There is also a small but non-negligible audience of active followers in the capital, Port Talbuk.

The central theme of DLB propaganda has been to reject the corrupting influence of Western capitalism by disrupting the flow of commerce through the Straits of Arpanda while maintaining and expanding the group's current holdings in Takanwei. DLB propaganda includes six lines of effort:

- recruiting (fighters, patriotic hackers, and general supporters on social media)
- threats (to shipping, town councils of elders, and citizens in areas the DLB wishes to control)
- attacks on legitimacy (of GOKA, GOT, and MNF-DP)
- gaining community support (often coordinated with face-to-face outreach in an effort to balance the diminution of support from collateral damage)
- shifting blame for anything and everything to GOKA
- promoting or calling for an autonomous region in southern Takanwei "free from government repression and foreign interference."

Specific acts of propaganda include various threats to shipping transiting the straits, including threatening messages and the detonation of devices in and near the straits, a threat video in June 2022 showing the release of maritime mines in the straits, and attacks from Q-ships. Other activities have included threats to the council of elders in the southwestern coastal town of Blackshores and the March 2022 execution of several men in Port Cook who refused to join the DLB. The group's anti-Western rhetoric includes abundant false "facts" and fallacious arguments focusing on perceived unfair economic and social policies aimed at the region's people; Western support of the wealthy, corrupt regional governments and political parties; and economic imperialism over the poor in the region.

### Road to Crisis

In January 2022, DLB aggression accelerated, beginning the road to crisis. The UN issued its resolution approving MNF-DP in June 2022, with operations beginning in August of that year. Appendix B provides a month-by-month summary of key events on the road to crisis.

## Relevant Elements of the MNF-DP Campaign

*We've been planning around the clock since we received the warning order that anticipated the possibility of this operation. I've worked hard to make sure that the cognitive and informational aspects of the operation received due attention in the overall plan. I also know that* **effective assessment starts in the planning phase**, *so I've been making sure that the guidance generated through the staff's operational design process leads to specific objectives that we can assess against. I've also been making sure that we can identify and insert points in the plan where we need to be collecting data to enable assessments, as well as points in the plan where we need to make decisions informed by assessments. After all,* **assessment must support decisionmaking***.*

The JTF headquarters leading MNF-DP produced the following as part of commander's guidance supporting subordinate detailed planning. This guidance includes the mission statement, commander's intent, the commander's desired end state, and the IO concept of operations, as well as a summary of lines of operation within that concept of operations.

### Mission Statement

> "Multi National Force–Dunarian Peninsula (MNF-DP), led by the United States, will assist Arpanda and Takanwei in neutralizing the Dunarian Liberation Brotherhood (DLB) and restoring secure and stable conditions at a level that each host nation can sustain."

## Commander's Intent

Protect and maintain freedom of transit of the Straits of Arpanda, coordinate with Arpanda and neighboring countries to neutralize the DLB and its asymmetrical capabilities by integrating and synchronizing information operations (IO) and information-related capabilities (IRCs) with maneuver forces.

## Commander's Desired End State

Key DLB sites seized and neutralized, DLB threat activities nullified, and collateral damage minimized. Security and stability in Arpanda and Takanwei returned to pre–January 2022 levels or better.

## IO Concept of Operations

MNF-DP, in coordination with U.S. Eastern Command and the governments of Arpanda and Takanwei, will execute the full integration of IO and IRCs to neutralize the DLB insurgency's asymmetrical activities. MNF-DP will conduct an overall IO campaign to shape the battlespace, thwart adversary efforts by garnering popular support in regional partner nations, and apply IRCs across all levels of warfare and phases of operation to protect access to the Straits of Arpanda. MNF-DP commander authorizes the execution of those IO tasks and other informational tasks derived from the MNF-DP J2 analysis of the operational environment.

## Lines of Operation

To counter DLB efforts to disrupt shipping in the Straits of Arpanda, MNF-DP will assist the governments of Arpanda, Takanwei, and other nations by logically employing IO tactics and IRCs in support of five lines of operation:

1. MNF-DP integrates IO with sea, air, and land forces to neutralize DLB C2 [command-and-control] networks (CYBER/EW [electronic warfare]/MISO/PA [public affairs]).

2. MNF-DP employs IO and IRCs to deny safe havens to DLB (CYBER/MISO/PA).

3. MNF-DP nullifies DLB IO and asymmetrical capabilities (CYBER/MISO/PA/SPACE).

4. MNF-DP utilizes IO to support stability operations in the region (MISO/PA/CA).

5. Promote economic development and good governance (MISO/PA/CA).

*Once the staff had agreed on the primary lines of operation for IO, my IO working group (IOWG) was able to generate IO objectives and subordinate objectives, which we could then support with specific IRC executions. Identifying subordinate objectives made it easier to identify and plan specific IRC executions, and it will also help with assessment by **breaking objectives into smaller subordinate objectives or sequential steps**. Laying out the subordinate objectives and thinking about how they would contribute to the larger objectives also served as a first step in laying out **the logic of the effort (or theory of change) connecting our intended activities with our objectives**.*

## IO Objectives

The following IO objectives support these lines of operation:

- Objective 1. Deny safe havens to DLB/isolate DLB from the populace.
- Objective 2. Counter DLB propaganda.
- Objective 3. Disrupt/degrade DLB C2.
- Objective 4. Protect MNF-DP movement and friendly CA/civil-military operations (CMO).
- Objective 5. Promote economic development and good governance.

In turn, each of these five objectives is supported by a number of subordinate objectives. Subordinate objectives are listed only for IO objectives 1 and 2, as these are the objectives that MAJ Fnorky will emphasize in the worked example in Chapter Four:[1]

- Objective 1. Deny safe havens to DLB/isolate DLB from the populace.
  - 1a. Deny DLB access to new areas/safe havens.
    - 1a1. Physically interdict DLB access to new safe havens.
    - 1a2. Populace rejects DLB to prevent access to new safe havens.
  - 1b. Push DLB out from existing safe havens.
    - 1b1. Decrease DLB freedom of movement within existing safe havens.
    - 1b2. Populace pushes DLB out of current safe havens.
    - 1b3. Increase action against DLB (e.g., through arrest, popular pressure) in existing safe havens.
    - 1b4. Increase security force presence in existing safe havens.

---

[1]  While objectives 3, 4, and 5 are clearly within (or supported by) the IO portfolio, they are not specifically IIP objectives—the focus of this series of reports. Assessment for objectives 3 and 4 should be relatively straightforward. Assessment for objective 5 is both complicated and long-term. While the assessment principles espoused here are certainly applicable to efforts to assess development and governance, there is a substantial literature focused more explicitly on that topic. See, for example, Jan Osburg, Christopher Paul, Lisa Saum-Manning, Dan Madden, and Leslie Adrienne Payne, *Assessing Locally Focused Stability Operations*, Santa Monica, Calif.: RAND Corporation, RR-387-A, 2014.

- ○ 1b5. Decrease support to/cooperation with DLB from transnational criminal organizations (TCOs) to push DLB out of existing safe havens.

- Objective 2. Counter DLB propaganda.
  - – 2a. Reduce DLB message transmission/reach.
    - ○ 2a1. Destroy or disable DLB propaganda transmission means.
    - ○ 2a2. Reduce DLB propaganda producers' freedom of action.
    - ○ 2a3. Usurp DLB propaganda sources/means and replace with favorable messaging.
  - – 2b. Refute false DLB claims.
    - ○ 2b1. Broadcast true information that contradicts DLB claims.
    - ○ 2b2. Demonstrate the falsehood of DLB claims.
  - – 2c. Counter the effects of DLB propaganda.
    - ○ 2c1. Inoculate target audiences against DLB messages.
    - ○ 2c2. Undermine the credibility of DLB propaganda sources and producers.
    - ○ 2c3. Reassure/promote confidence in GOKA, GOT, MNF-DP, and the security of the straits.

*As we worked to refine objectives and to plan integrated IRC executions to meet these objectives, we continued to think about how we would assess those executions and progress toward the objectives. Thinking about the two together was really productive—especially thinking about the **logic of the effort/theory of change** underlying the actions we were considering. As we brainstormed ideas for IRC actions, we tried to plug them into our evolving understanding of the logic of the effort. This got easier as we started to **break objectives into smaller subordinate objectives**, because more granular objectives helped us think of ways we might employ IRCs to lead to those things! With our planned actions clearly and logically connected to our assumptions about how they would lead to the accomplishment of our objectives, I knew we were preparing ourselves for an easier time planning and conducting assessment.*

**Planned IRC Tasks**

Objectives 1 and 2 include five intermediate objectives and 15 subordinate objectives. Sixteen IRC tasks support these subordinate objectives. Most of the 16 tasks support more than one subordinate objective, and most of the subordinate objectives are supported by more than one task. Table 3.1 summarizes the relationships between these subordinate objectives and tasks. Because most of the planned tasks or actions support more than one objective or subordinate objective, every task has a unique number, and that number remains the same each time it is listed (so, for example, subordinate objective 1b5 is supported by tasks 6, 7, and 4, where 6 and 7 are newly listed but 4 has been listed in support of earlier subordinate objectives).

**Table 3.1**
**Crosswalk of Subordinate Objectives and Planned IRC Tasks**

Subordinate Objectives (columns):

- 1a2. Populace rejects DLB to prevent access to new safe havens.
- 1b1. Decrease DLB freedom of movement within existing safe havens.
- 1b2. Populace pushes DLB out of current safe havens.
- 1b3. Increase action against DLB (e.g., through arrest, popular pressure) in existing safe havens.
- 1b5. Decrease support to/cooperation with DLB from transnational TCOs to push DLB out of existing safe havens.
- 2a1. Destroy or disable DLB propaganda transmission means.
- 2a2. Reduce DLB propaganda producers' freedom of action.
- 2a3. Usurp DLB propaganda sources/means and replace with favorable messaging.
- 2b1. Broadcast true information that contradicts DLB claims.
- 2b2. Demonstrate the falsehood of DLB claims.
- 2c1. Inoculate target audiences against DLB messages.
- 2c2. Undermine the credibility of DLB propaganda sources and producers.
- 2c3. Reassure/promote confidence in GOKA, GOT, MNF-DP, and the security of the straits.

| Task | 1a2 | 1b1 | 1b2 | 1b3 | 1b5 | 2a1 | 2a2 | 2a3 | 2b1 | 2b2 | 2c1 | 2c2 | 2c3 |
|---|---|---|---|---|---|---|---|---|---|---|---|---|---|
| 1. Publicize MNF-DP and Takanwei PSF efforts to patrol, establish checkpoints, arrest, and combat DLB in pursuit of subordinate objective 1a1 (PA, MISO). | | ▓ | | ▓ | | | | | | ▓ | | | ▓ |
| 2. Encourage local populations in areas at risk to vocally and physically reject DLB incursions (MISO). | ▓ | | ▓ | | | | | | | | | | |
| 3. Encourage local populations in areas at risk to report DLB presence and activities to a tip line (MISO). | | ▓ | ▓ | ▓ | | | | | | | | | |
| 4. Reinforce messages with action (maneuver forces, CMO). | | | | | ▓ | | | | ▓ | ▓ | | | ▓ |
| 5. Advertise CMO/security force successes, emphasizing success of cooperation and legitimacy of GOT and MNF-DP (PA, MISO). | | | | | | | | | ▓ | | ▓ | | |
| 6. Advertise relationship between DLB and TCOs, and demonize both (PA, MISO). | | | | | ▓ | | | | | | ▓ | ▓ | |
| 7. Signal to TCOs a willingness to reduce attention to their operations if they reduce cooperation with DLB (MISO, key leader engagement). | | | | | ▓ | | | | | | | | |

**Table 3.1—Continued**

| Task | Subordinate Objective | | | | | | | | | | | | |
|---|---|---|---|---|---|---|---|---|---|---|---|---|---|
| | 1a2. Populace rejects DLB to prevent access to new safe havens. | 1b1. Decrease DLB freedom of movement within existing safe havens. | 1b2. Populace pushes DLB out of current safe havens. | 1b3. Increase action against DLB (e.g., through arrest, popular pressure) in existing safe havens. | 1b5. Decrease support to/cooperation with DLB from transnational TCOs to push DLB out of existing safe havens. | 2a1. Destroy or disable DLB propaganda transmission means. | 2a2. Reduce DLB propaganda producers' freedom of action. | 2a3. Usurp DLB propaganda sources/means and replace with favorable messaging. | 2b1. Broadcast true information that contradicts DLB claims. | 2b2. Demonstrate the falsehood of DLB claims. | 2c1. Inoculate target audiences against DLB messages. | 2c2. Undermine the credibility of DLB propaganda sources and producers. | 2c3. Reassure/promote confidence in GOKA, GOT, MNF-DP, and the security of the straits. |
| 8. Pursue terms-of-service violations by DLB social media accounts, suspend accounts at the university, etc. (Judge Advocate General). | | | | | | ▓ | ▓ | | | | | | |
| 9. Destroy or disable transmission means (EW, offensive cyber operations, physical attack). | | | | | | ▓ | | | | | | | |
| 10. Threaten identified DLB propaganda producers with exposure, arrest, or death (MISO). | | | | | | | ▓ | | | | | | |
| 11. Target (for arrest, capture, or strike) DLB propaganda producers (law enforcement, physical attack). | | | | | | | ▓ | | | | | | |
| 12. Usurp or overwhelm DLB channels/social media (cyber, EW). | | | | | | | | ▓ | | | | | |
| 13. Broadcast on usurped channels/means, impersonating DLB representatives either to promote harmless action with an opportunity for MNF-DP response (MILDEC [military deception]) or to broadcast MISO messages relevant to other objectives and subordinate objectives (MISO). | | | | | | | | ▓ | | | ▓ | ▓ | |

**Table 3.1—Continued**

| Task | Subordinate Objective | | | | | | | | | | | | |
|------|------|------|------|------|------|------|------|------|------|------|------|------|------|
| | 1a2. Populace rejects DLB to prevent access to new safe havens. | 1b1. Decrease DLB freedom of movement within existing safe havens. | 1b2. Populace pushes DLB out of current safe havens. | 1b3. Increase action against DLB (e.g., through arrest, popular pressure) in existing safe havens. | 1b5. Decrease support to/cooperation with DLB from transnational TCOs to push DLB out of existing safe havens. | 2a1. Destroy or disable DLB propaganda transmission means. | 2a2. Reduce DLB propaganda producers' freedom of action. | 2a3. Usurp DLB propaganda sources/means and replace with favorable messaging. | 2b1. Broadcast true information that contradicts DLB claims. | 2b2. Demonstrate the falsehood of DLB claims. | 2c1. Inoculate target audiences against DLB messages. | 2c2. Undermine the credibility of DLB propaganda sources and producers. | 2c3. Reassure/promote confidence in GOKA, GOT, MNF-DP, and the security of the straits. |
| 14. Monitor DLB claims and selectively refute false claims (where doing so will not generate more attention for DLB than its original false claims did) through government and MNF-DP media (MISO) and press releases (PA). | | | | | | | | | ▓ | | | ▓ | |
| 15. Using intelligence on planned new DLB propaganda themes, identify actions to take and messages to broadcast that will undercut those new themes (intel, maneuver, MISO). | | | | | | | | | | | ▓ | | |
| 16. Use intelligence to identify and broadcast embarrassing or counterproductive information from the personal history of individual DLB propagandists or from recordings of their recent utterances or actions (intel, MISO). | | | | | | | | | | | | ▓ | |

NOTE: Subordinate objectives 1a1 ("Physically interdict DLB access to new safe havens") and 1b4 ("Increase security force presence in existing safe havens") are excluded because they are exclusively kinetic objectives.

The following tasks support subordinate objective 1a2, "Populace rejects DLB to deny access to new safe havens":

1. Publicize MNF-DP and Takanwei PSF efforts to patrol, establish checkpoints, arrest, and combat DLB in pursuit of subordinate objective 1a1 (PA, MISO).
2. Encourage local populations in areas at risk to vocally and physically reject DLB incursions (MISO).
3. Encourage local populations in areas at risk to report DLB presence and activities to a tip line (MISO).
4. Reinforce messages with action (maneuver forces, CMO).
5. Advertise CMO/security force successes, emphasizing success of cooperation and legitimacy of GOT and MNF-DP (PA, MISO).

The following tasks support subordinate objective 1b1, "Decrease DLB freedom of movement within existing safe havens":

3. Encourage local populations to report DLB presence and activities to a tip line (MISO).
4. Reinforce messages with action (maneuver forces, CMO).

The following tasks support subordinate objective 1b2, "Populace pushes DLB out of existing safe havens":

1. Publicize MNF-DP and Takanwei PSF efforts to patrol, establish checkpoints, arrest, and combat DLB in pursuit of subordinate objective 1b1 (PA, MISO).
2. Encourage local populations in areas at risk to vocally and physically reject DLB incursions (MISO).
3. Encourage local populations in areas at risk to report DLB presence and activities to a tip line (MISO).
4. Reinforce messages with action (maneuver forces, CMO).
5. Advertise CMO/security force successes in pursuit of subordinate objectives 1b1 and 1b3, emphasizing success of cooperation and legitimacy of GOT and MNF-DP (PA, MISO).

The following tasks support subordinate objective 1b3, "Increase action against DLB in existing safe havens":

3. Encourage local populations in areas at risk to report DLB presence and activities to a tip line (MISO).
4. Reinforce messages with action (maneuver forces, CMO).

The following tasks support subordinate objective 1b5, "Decrease support to/cooperation with DLB from TCOs to push DLB out of existing safe havens":

4. Reinforce messages with action (maneuver forces, CMO).
6. Advertise relationship between DLB and TCOs, and demonize both (PA, MISO).
7. Signal to TCOs a willingness to reduce attention to their operations if they reduce cooperation with DLB (MISO, key leader engagement).

The following tasks support subordinate objective 2a1, "Destroy or disable DLB propaganda transmission means":

8. Pursue terms-of-service violations by DLB social media accounts, suspend accounts at the university, etc. (Judge Advocate General).
9. Destroy or disable transmission means (EW, offensive cyber operations, physical attack).

The following tasks support subordinate objective 2a2, "Reduce DLB propaganda producers' freedom of action":

8. Pursue terms-of-service violations by DLB social media accounts, suspend accounts at the university, etc. (Judge Advocate General, CA).
10. Threaten identified DLB propaganda producers with exposure, arrest, or death (MISO).
11. Target (for arrest, capture, or strike) DLB propaganda producers (law enforcement, physical attack).

The following tasks support subordinate objective 2a3, "Usurp DLB propaganda sources/means and replace with favorable messaging":

12. Usurp or overwhelm DLB channels/social media (cyber, EW).
13. Broadcast on usurped channels/means, impersonating DLB representatives either to promote harmless action with an opportunity for MNF-DP response (MILDEC) or to broadcast MISO messages relevant to other objectives and subordinate objectives (MISO).

The following tasks support of subordinate objective 2b1, "Broadcast true information that contradicts DLB claims":
5. Advertise CMO/security force successes, emphasizing success of cooperation and legitimacy of GOT and MNF-DP (PA, MISO).
6. Advertise relationship between DLB and TCOs, and demonize both (PA, MISO).

14. Monitor DLB claims and selectively refute false claims (where doing so will not generate more attention for DLB than its original false claims did) through government and MNF-DP media (MISO) and press releases (PA).

The following tasks support subordinate objective 2b2, demonstrate falsehood of DLB claims:

1. Publicize MNF-DP and Takanwei PSF efforts to patrol, establish checkpoints, arrest, and combat DLB (PA, MISO).
4. Reinforce messages with action (maneuver forces, CMO).
5. Advertise CMO/security force successes, emphasizing success of cooperation and legitimacy of GOT and MNF-DP (PA, MISO).

The following tasks support subordinate objective 2c1, inoculate target audiences against DLB propaganda messages:

5. Advertise CMO/security force successes, emphasizing success of cooperation and legitimacy of GOT and MNF-DP (PA, MISO).
6. Advertise relationship between DLB and TCOs, and demonize both (PA, MISO).
13. Broadcast through usurped channels/means, impersonating DLB representatives either to promote harmless action with an opportunity for MNF-DP response (MILDEC) or to broadcast MISO messages relevant to other objectives and subordinate objectives (MISO).
15. Using intelligence on planned new DLB propaganda themes, identify actions to take and messages to broadcast that will undercut those new themes (intel, maneuver, MISO).

The following tasks support of subordinate objective 2c2, "Undermine the credibility of DLB propaganda sources and producers":

6. Advertise relationship between DLB and TCOs, demonize both (PA, MISO).
13. Broadcast through usurped channels/means, impersonating DLB representatives either to promote harmless action with an opportunity for MNF-DP response (MILDEC) or to broadcast MISO messages relevant to other objectives and subordinate objectives (MISO).
14. Monitor DLB claims and selectively refute false claims (where doing so will not generate more attention for DLB than its original false claims did) through government and MNF-DP media (MISO) and press releases (PA).
16. Use intelligence to identify and broadcast embarrassing or counterproductive information from the personal history of individual DLB propagandists or from recordings of their recent utterances or actions (intel, MISO).

The following tasks support subordinate objective 2c3, "Reassure/promote confidence in GOKA, GOT, MNF-DP, and the security of the straits":

1. Publicize MNF-DP and Takanwei PSF efforts to patrol, establish checkpoints, arrest, and combat DLB (PA, MISO).
5. Advertise CMO/security force successes, emphasizing success of cooperation and legitimacy of GOT and MNF-DP (PA, MISO).

*So, we ended up with plans for 16 IRC tasks or efforts to support two numbered objectives, five numbered and lettered intermediate objectives, and 15 numbered, lettered, and numbered subordinate objectives. As part of our assessment effort, we'll need to monitor all 16 tasks and assess progress toward all 22 nested objectives and subordinate objectives. As noted, laying out the objectives in a hierarchy and breaking out the specific tasks goes a long way toward articulating our* **logic of the effort/theory of change**. *The next step is assessment planning, because* **effective assessment starts in the planning phase**. *I will not be surprised if something that comes up in assessment planning causes a change in our thinking that feeds back and makes us change one or more of our objectives (or otherwise change something in the plan). And, that is okay, because I know those changes will be* <u>*improvements*</u>.

# CHAPTER FOUR

# Worked Example of Assessment Design and Planning for Selected Campaign Elements

This chapter follows MAJ Fnorky as he works through assessment planning, starting with preliminary IO objectives 1 and 2 (and their five intermediate objectives and 15 subordinate objectives), as well as the 16 preliminary planned IRC tasks that support them.

*Right! Assessment planning. Where to begin? Well, the IO working group is **starting its assessment planning as part of overall campaign planning**, so we've got that part right. I guess going forward we'll work through the big takeways for assessment that I distilled from my reading. First up is **effective assessment requires clear, realistic, and measurable goals**. We put a lot of thought and effort into identifying our goals, and we're clear on how they connect to broader mission objectives and how the tasks and actions we're planning will help us achieve those goals. But let's think more about those objectives from an assessment perspective.*

## Effective Assessment Requires Clear, Realistic, and Measurable Goals

*You can't assess against something if you can't tell whether or not you have accomplished it. Requirements for objectives need to be SMART: specific, measurable, achievable, relevant, and time-bound. Let's roll through our objectives and subordinate objectives and see whether they really are SMART. . . .*

### Objective 1: Deny Safe Havens to DLB/Isolate DLB from the Populace

Although objective 1, as worded, is sufficient to begin thinking about a theory of change and identifying general IRC efforts that might support its accomplishment, it is not good enough (by itself) for assessment. It is not really good enough for detailed planning, either. Objective 1 lacks details: It does not specify who or what will be involved in achieving it when and where it will be achieved. Use the SMART criteria to refine your objectives.

### Specific

Objective 1, as written, is more of a headline or summary statement: It really lacks specificity. *Safe havens* is not specific. Which areas, exactly? How are they defined, or bounded? *Populace* is not specific, either. Which populace? Does this include everyone within the areas defined as safe havens? Everyone in Takanwei? Which *specific* population segments in which areas? *Deny* is probably sufficiently specific (there is an understanding in military parlance of what constitutes denial of an area), but *isolate* is probably not. *Isolate* could mean many things: preventing an adversary from achieving physical proximity, preventing it from gathering support (intelligence, supplies/resources, or both), or psychological isolation.

Furthermore, there is a lack of specificity about the level of achievement required (it is good to have **target thresholds** and a clear idea of **how much is enough**). Is the threshold complete denial/isolation? A 50-percent reduction in DLB presence? A 50-percent reduction in the number of safe havens in use? And what are expectations for how long this will take? Will the effect be seen immediately? Within a week? Within six months? Perhaps the best approach is to identify a range of thresholds—**broken into incremental and sequential steps**—and to tie these increments to phases of the operation. For example, you might expect a 10-percent reduction in DSB presence in a particular village by the end of phase I of the operation, a cumulative 25-percent reduction by the end of phase II, and so on.

When setting target thresholds, it can help to think about how much the overall success of the operation hinges on accomplishing this objective and what level of accomplishment will have a positive (or sufficient) operational impact. This process can help you set a minimum target threshold, which is a far better basis for assessment than a desired or intended level of accomplishment. **Remember, good objectives need to at least imply what failure would look like**.

### Measurable

The more specific objectives get, the more measurable they become. With additional specificity discussed in the previous section, objective 1 looks like a measurable objective. Whatever intelligence collection means you have used to identify DLB safe havens should also be able to tell you if and when the group stops using those safe havens.[1] Furthermore, it is good that you have access to intelligence reports on the location of DLB safe havens prior to the commencement of operations, because **evaluating change requires a baseline**. DLB presence is something that is observable, even if it may take considerable effort to arrange for the collection of those observations. So, denial of safe havens is measurable, at least in principle.

---

[1]   The J2 section (intelligence) is always heavily tasked and usually receives more requests for information than it can respond to. Coordination with J2 is key. This coordination will be easiest when you are able to request intelligence that J2 is *already collecting* to inform your assessments. Requests that require new data collection (or even new reporting formats) will compete for priority with other new requests from other staff sections.

*Isolation* might also be measurable, depending on how it is specified. Physical proximity is at least notionally observable, but that is probably what is really desired here. Psychological isolation might be measurable with various psychographic surveys and instruments, so that is at least notionally measurable, too. However, it is pretty unlikely that you can get DLB members to allow you to measure their feelings of isolation. Therefore, psychological isolation is probably a poor choice for objective refinement, though it might be possible to come up with indicators that could be collected without communicating directly with DLB members.

The overall focus of objective 1 is on denying sanctuary and resources to the DLB, so perhaps the best specification of *isolate* would be to prevent the DLB from gathering tangible support (e.g., intelligence, supplies) from the populace (however the populace is defined). Not only does this specification connect to the concept of the objective, but it can also help **specify the observable behaviors sought and from whom they are sought**. The objective is to prevent the DLB from being able to take the things they need from the populace and to prevent the populace from *giving* the DLB tangible support. Is this measurable? Again, it is notionally observable, though it may require some effort to get observations.

### Achievable

The extent to which objective 1 is achievable is entirely dependent on the specifics of the level of accomplishment and timeline. Can the DLB be completely denied access to safe havens and completely isolated from the population in less than a week? Absolutely not! Given infinite time and resources, could the DLB be denied and isolated? Almost assuredly. Of course, the resources and time are never infinite. How, then, can one discern whether an objective is achievable? This is where formative assessment is extremely valuable.[2] Consider previous efforts directed at this aspect of counterinsurgency. How long have they taken? What level of resources have they required, and can you scale that to the size of Takanwei and the DLB? Now, consider prior performance and the level of available resources. Do they, collectively, suggest reasonable prospects for achieving the objective? If not, the target thresholds may need to be reduced, the planned resources may need to be increased, or a more fundamental change to the plan may be required.

### Relevant

Does objective 1 actually support broader operational objectives? That is, does objective 1 actually *nest* with and contribute to broader operational goals? Sure it does. The mission statement sets the clear goal of neutralizing the DLB, and it logically follows that denying the group sanctuary and reducing the tangible support it receives from the populace will make it harder for the DLB to operate and reduce its freedom of

---

[2]  For a detailed discussion of formative assessment, see Chapter Seven in the desk reference (Paul, Yeats, Clarke, and Matthews, 2014).

movement, enabling other efforts to neutralize it. It is always good to double-check an objective's relevance to make sure that all goals and subordinate goals logically nest with each other. If so, achieving subordinate objectives will clear a direct path to achieving higher-level objectives.

### Time-Bound

As written, objective 1 is *not* time-bound. There is no explicit start or end time, no explicit point at which the objective becomes overtaken by events and is no longer worth pursuing. But, as discussed under "Specific," that is something that should be specified. The objective should be refined to include sequential, incremental targets, which should be tied to a timeline—perhaps a flexible timeline—like the phases of the operational plan.

*So, objective 1, "Deny safe havens to DLB/isolate DLB from the populace," is a pretty good summary of an objective, but—by itself—it isn't SMART. To get it to SMART, we probably need an objective paragraph to follow the headline. In that elaboration, we should make clear that we mean* **existing safe havens** *(and we should specify and bound them), as well as* **potential new safe havens** *anywhere in the south of Takanwei. We'll specify that isolation refers to cutting off the provision of tangible support (that is, intelligence, supplies, food, materiel, etc.) to the DLB by the populace. We'll specify the* **populace** *as relevant segments of supporting populations with the understanding that we still need to be more specific and do some target audience analysis (or leverage other information) to establish a baseline for who is currently providing support to the DLB and what kind of support they are providing. We'll also establish target thresholds for* **denial and isolation**, *and we'll tie incremental thresholds to the phases of the operation. The "S" in SMART is for specific, and most of what we're adding to the objective statement is—in some form or another—specificity. I guess the -MART part is mostly a reminder of other things we need to be specific about!*

### Objective 2: Counter DLB Propaganda

Objective 2 is "Counter DLB propaganda." Again, this is a good place to start and a reasonable high-level objective, but it is not good enough for assessment or planning. Use the SMART criteria to refine this objective.

### Specific

Objective 2 really needs to be unpacked and specified. Like objective 1, its key terms are underspecified. What is intended by *counter*? By how much and by when? The solution used in objective 1—tying **sequential and incremental target thresholds** to operational phases—should work here, too. As was probably the case for objective 1, looking at the intermediate objectives for objective 2 helps specify *counter*. Intermediate objective 2a is "Reduce DLB message transmission/reach," so, counter the propaganda by reducing the amount of it, or reduce the area in which it is available.

Intermediate objective 2b is "Refute false DLB claims," which suggests countering the propaganda by discrediting it.

Intermediate objective 2c is "Counter the effects of DLB propaganda," which begs the questions of what effects DLB propaganda is trying to cause and what effects it is actually having. Presumably, the effects of DLB propaganda can be described in terms of **specific observable behaviors sought** (such as joining the DLB, giving the DLB money, joining a protest, or even just following the DLB on social media) and **from whom they are sought** (specific target audiences). It follows that countering the effects of DLB propaganda can be specified as (1) no change in these behaviors among the targeted groups or (2) diminished frequency of these behaviors among the targeted groups. All three intermediate objectives, and especially the third one, necessitate good baseline measurement of the current effort and its effects because **evaluating change requires a baseline**.

### Measurable

As written, objective 2 is not really measurable, but it can be measurable if it is articulated with greater specificity. Suppose you specify *counter* as reducing transmission/reach, refuting false claims, and countering effects. All of these things can be measured.

### Achievable

Like objective 1, the extent to which objective 2 is achievable depends on the specific target thresholds set and on the resources dedicated to pursuing the objective. The DLB does not have a particularly robust or extensive propaganda apparatus, and this apparatus has limited redundancy, so a modest level of success against DLB propaganda appears to be very achievable.

### Relevant

How well does objective 2 support the mission of neutralizing the DLB? The DLB uses propaganda to recruit, promote support, and make threats (with operational consequences), so countering DLB propaganda is certainly relevant to the mission. How about intermediate objectives—the ones used to specify what *counter* means here? The connection between reducing transmission and reach (intermediate objective 2a) and reducing the impact of DLB propaganda is clear, as is the connection between countering the effects of DLB propaganda (intermediate objective 2c) and neutralizing the DLB.

What about intermediate objective 2b, "Refute false DLB claims"? The relevance of this objective is based on assumptions that false DLB claims have an effect and that refuting those false claims will diminish that effect. This is certainly a plausible set of assumptions, but perhaps they merit more attention. (This will be revisited in the discussion that follows on theory of change/logic of the effort.)[3]

### Time-Bound

Like objective 1, as written, objective 2 is not time-bound. If incremental thresholds are specified and linked to a timeline or to phases of the larger operation, objective 2 becomes time-bound.

The model used for the SMART review for objectives 1 and 2 should be followed for all objectives, intermediate objectives, and subordinate objectives. Specifying the higher-level objectives makes it easier to SMART up the intermediate and subordinate objectives and to make sure they fully nest below the higher-level objectives. Due to space constraints, this report does not include a detailed review of the 15 subordinate objectives under objectives 1 and 2, but it refers to them as needed in the remainder of the worked example.

*It took us a while, but we finally got through all the objectives and intermediate and subordinate objectives. It was a grind at first, but then it got easier. It got easier for two reasons. First, we just got better at it as we got more practice. Second, as we got deeper into the nested objectives, we found that we had already done the hard thinking when we were reviewing the parent objective, so we just needed to apply that thinking again. I bet the next time we do this our initial objectives will start out a lot closer to SMART!*

## Effective Assessment Requires a Theory of Change/Logic of the Effort Connecting Activities to Objectives

*The **importance of a theory of change or logic of the effort** has been stuck in my mind since I encountered the idea in my reading. We thought about the logical connections between activities and objectives when we were originally planning objectives and tasks, and the logic of the effort came up in our thinking every time we got to **relevant** in our SMART review: Did the objective under consideration actually logically support the objective it was supposed to be nested under? I knew all of that was building up to formally laying out our theory of change (or theories, since there is more than one logic here), so I think that'll go fairly smoothly. And I bet we'll gain some useful insights from doing it.*

The following discussion walks through the assembly of a logic model for objective 1. The discussion proceeds step by step, illustrating incomplete or in-progress logic models before the final model is presented in Table 4.3. The model is built "back-

---

[3]   For a discussion of the effectiveness (or, actually, the ineffectiveness) of trying to directly refute false propaganda, see Christopher Paul and Miriam Matthews, *The Russian "Firehose of Falsehood" Propaganda Model: Why It Might Work and Options to Counter It*, Santa Monica, Calif.: RAND Corporation, PE-198-OSD, 2016.

ward," starting with the tiered objectives as tiered outcomes. The implication is that all the subordinate outcomes will add up to the desired ultimate outcome: accomplishing objective 1. These outcomes are listed out in Table 4.1, culminating on the top right with the highest-level objective in the model, objective 1, with everything that contributes to it building up from the left.

Table 4.1 captures part of what has already been thought through in planning and in the SMART review: how the various objectives nest logically with each other and build to the highest-level objective. The second step in this kind of backward logic modeling is to finish adding in the work that has already been done. That is, add in the numbered tasks supporting the corresponding objectives. Table 4.2 presents this second step in logic model building, with the addition of the identified tasks as "activities."

The third step toward completing the logic model involves noting additional activities and specifying the needed inputs and outputs. In this case, some activities require capabilities outside the traditional core IRCs, such as actions from fire or maneuver elements. These activities were excluded from the IO-specific task list, even though they were explicit in the objectives. They can easily be spelled out here, however. Traditional logic models list a wide range of things as inputs—usually resources of different kinds (e.g., money, personnel, transportation)—at levels of detail that depend on the purpose. Here, it is probably sufficient to list the IRCs or other capabilities expected to execute the activity. When reviewing the logic model after it is complete, one of the standard checks is to make sure that the needed inputs will be available. Rather than listing specific resources and checking availability, this review will also need to consider whether the identified IRCs or capabilities would typically have the means available to execute the specified activity.

*As we finished up the initial draft of the logic model for objective 1 by listing the capability inputs and the desired outputs, we started to carefully go through and make sure everything made sense logically. One of the IOWG members noticed something odd about subordinate objective 1b3, "Increase action against DLB in existing safe havens." He pointed out that this isn't really an* **outcome***, but it is an* **output***. It is in the wrong column. It still fits logically, but it should be an output supporting objective 1b1, "Decrease DLB freedom of movement within existing safe havens." He's right. Looking over the other subordinate objectives, we realized that 1b4 ("Increase security force presence in safe havens") is also an output tied to subordinate objective 1b1. With these moved one column to the left, things made a little more sense. These two changes are called out with bold text in Table 4.3, and also outlined in red.*

Table 4.3 presents a complete basic logic model. For the purposes of this worked example, this logic model is just about the right size and level of detail. The next step is to **identify and watch for possible constraints, barriers, disruptors, and unintended consequences**. To do this, you will review the logic model to make sure it makes sense and to identify things that might prevent planned efforts from succeeding. Such a review should consider each column of the logic model as well as the connections between the columns. What assumptions are being made? Are those assumptions reasonable or are they questionable? Are available forces really going to be able to perform the specified activities and produce needed outputs? What might adversaries do that could (intentionally or otherwise) interfere with the effort?

By identifying bad linkages, vulnerable assumptions, and ways in which the adversary might disrupt actions *during the planning phase*, you can adjust plans early on to make them less vulnerable, or you can monitor vulnerable aspects of an effort and know immediately if things go awry. Corrections can be made in the field if an assumption does not pan out or if an adversary action interferes. But this can happen only *if we are aware that the chain of logic has been broken*.

Examining the logic model in Table 4.3 for weaknesses or vulnerabilities might yield several points of possible concern. First, as is often the case with IIP campaigns, the whole line of effort is based on uncertain assumptions. The logic model for objective 1 basically assumes that increasing security and services, combined with increasing awareness of security and services and a demonstrated effectiveness against the DLB in the area, will encourage the specified populations to reject the DLB. While this sounds good, it is uncertain! It is plausible and follows a clear logic, but the logical connections are untested assumptions. This set of assumptions connects the outputs (all of which can likely be generated) with the desired first level of outcomes. We hope that these outputs will lead to these outcomes, but we are not completely confident that they will. And that is okay. By recognizing vulnerable assumptions, you will be able to plan for measurement collection that will allow you to monitor progress and validate or discard the assumptions as quickly as possible. When uncertain, **fail fast**.

An even more vulnerable set of assumptions (though perhaps less central to the core logic of the effort in pursuit of objective 1) supports subordinate objective 1b5, "Decrease support to/cooperation with DLB from TCOs to push DLB out of existing safe havens." Here, the core logic is that we can get the TCOs to reduce their support for the DLB either by negatively associating them with the DLB (making them reduce support to avoid bad press or because it will cost them support among local populations) or by threatening the TCOs with increased law-enforcement or MNF-DP attention if they do not reduce support to the DLB.

**Table 4.1**
**First Step in Building a Logic Model for Objective 1**

| Inputs | Activities | Outputs | Outcomes (subordinate objectives) | Outcomes (intermediate objectives) | Outcomes (objectives) |
|---|---|---|---|---|---|
| | | | 1a1. Physically interdict DLB access to new safe havens. | 1a. Deny DLB access to new areas/safe havens. | 1. Deny safe havens to DLB/isolate DLB from the populace. |
| | | | 1a2. Populace rejects DLB to prevent access to new safe havens. | | |
| | | | 1b1. Decrease DLB freedom of movement within existing safe havens. | 1b. Push DLB out from existing safe havens. | |
| | | | 1b2. Populace pushes DLB out of current safe havens. | | |
| | | | 1b3. Increase action against DLB (e.g., through arrest, popular pressure) in existing safe havens. | | |
| | | | 1b4. Increase security force presence in existing safe havens. | | |
| | | | 1b5. Decrease support to/cooperation with DLB from TCOs to push DLB out of existing safe havens. | | |

**Table 4.2**
**Second Step in Building a Logic Model for Objective 1**

| Inputs | Activities | Outputs | Outcomes (subordinate objectives) | Outcomes (intermediate objectives) | Outcomes (objectives) |
|---|---|---|---|---|---|
| | 1. Publicize MNF-DP and Takanwei PSF efforts to patrol, establish checkpoints, arrest, and combat DLB. | | 1a1. Physically interdict DLB access to new safe havens. | 1a. Deny DLB access to new areas/safe havens. | 1. Deny safe havens to DLB/isolate DLB from the populace. |
| | 2. Encourage local populations in areas at risk to vocally and physically reject DLB incursions (MISO). | | 1a2. Populace rejects DLB to prevent access to new safe havens. | | |
| | 3. Encourage local populations in areas at risk to report DLB presence and activities to a tip line. | | | | |
| | 4. Reinforce messages with action. | | | | |
| | 5. Advertise CMO/ security force successes, emphasizing success of cooperation and legitimacy of GOT and MNF-DP. | | | | |

**Table 4.2—Continued**

| Inputs | Activities | Outputs | Outcomes (subordinate objectives) | Outcomes (intermediate objectives) | Outcomes (objectives) |
|---|---|---|---|---|---|
| | 3 [repeated] <br> 4 [repeated] | | 1b1. Decrease DLB freedom of movement within existing safe havens. | 1b. Push DLB out from existing safe havens. | 1. Deny safe havens to DLB/isolate DLB from the populace (cont.). |
| | 1 [repeated] <br> 2 [repeated] <br> 3 [repeated] <br> 4 [repeated] <br> 5 [repeated] | | 1b2. Populace pushes DLB out of current safe havens. | | |
| | 3 [repeated] <br> 4 [repeated] | | 1b3. Increase action against DLB (e.g., through arrest, popular pressure) in existing safe havens. | | |
| | | | 1b4. Increase security force presence in existing safe havens. | | |
| | 4 [repeated] | | 1b5. Decrease support to/cooperation with DLB from TCOs to push DLB out of existing safe havens. | | |
| | 6. Advertise relationship between DLB and TCOs, and demonize both. | | | | |
| | 7. Signal to TCOs a willingness to reduce attention to their operations if they reduce cooperation with DLB | | | | |

**Table 4.3**
**Third (and Final) Step in Building a Logic Model for Objective 1, with a Key Change Highlighted**

| Inputs | Activities | Outputs | Outcomes (subordinate objectives) | Outcomes (intermediate objectives) | Outcomes (objectives) |
|---|---|---|---|---|---|
| Maneuver, police | Establish checkpoints, conduct patrols, surveillance | Checkpoints, patrols | 1a1. Physically interdict DLB access to new safe havens. | 1a. Deny DLB access to new areas/safe havens. | 1. Deny safe havens to DLB/isolate DLB from the populace. |
| PA, MISO | 1. Publicize MNF-DP and Takanwei PSF efforts to patrol, establish checkpoints, arrest, and combat DLB. | News stories, broadcasts; information received by specific populations | | | |
| MISO | 2. Encourage local populations in areas at risk to vocally and physically reject DLB incursions (MISO). | MISO products disseminated, received | 1a2. Populace rejects DLB to prevent access to new safe havens. | | |
| MISO | 3. Encourage local populations in areas at risk to report DLB presence and activities to a tip line. | MISO products disseminated, received; tip line established | | | |
| Maneuver forces, CMO | 4. Reinforce messages with action. | Arrests/captures in response to tips; improved provision of services | | | |
| PA, MISO | 5. Advertise CMO/ security force successes, emphasizing success of cooperation and legitimacy of GOT and MNF-DP. | News stories, broadcasts; information received by specific populations | | | |

Table 4.3—Continued

| Inputs | Activities | Outputs | Outcomes (subordinate objectives) | Outcomes (intermediate objectives) | Outcomes (objectives) |
|---|---|---|---|---|---|
| [repeated] | 3 [repeated]<br>4 [repeated] | **[repeated]; 1b3. Increase action against DLB (e.g., through arrest, popular pressure) in existing safe havens.** | 1b1. Decrease DLB freedom of movement within existing safe havens. | 1b. Push DLB out from existing safe havens. | 1. Deny safe havens to DLB/isolate DLB from the populace. |
| Maneuver, police | Establish checkpoints, conduct patrols, surveillance | **Checkpoints, patrolling; 1b4. Increase security force presence in existing safe havens.** | | | |
| [repeated] | 1 [repeated]<br>2 [repeated]<br>3 [repeated]<br>4 [repeated]<br>5 [repeated] | [repeated] | 1b2. Populace pushes DLB out of current safe havens. | | |
| [repeated] | 4 [repeated] | [repeated] | 1b5. Decrease support to/cooperation with DLB from TCOs to push DLB out of existing safe havens. | | |
| PA, MISO | 6. Advertise relationship between DLB and TCOs, and demonize both. | News stories, broadcasts; information received by specific populations | | | |
| MISO, key leader engagement | 7. Signal to TCOs a willingness to reduce attention to their operations if they reduce cooperation with DLB. | Messages sent to and received by TCO leadership | | | |

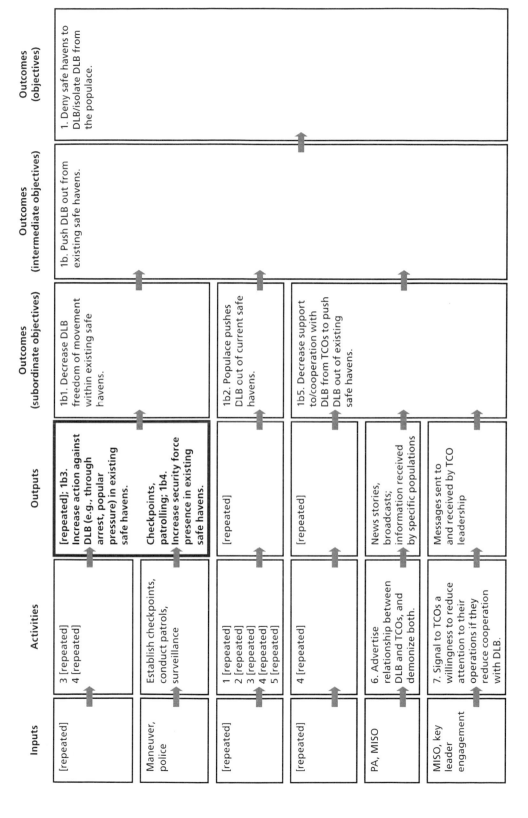

This logic is predicated on some fairly tenuous assumptions:

1. The TCOs care about being associated with the DLB, care about negative publicity, or care about the support of local populations.
2. The TCOs would acknowledge an increased threat from law enforcement or MNF-DP and believe those threats to be credible.
3. The TCOs support to the DLB is under the control of TCO leadership, and, if TCO leaders wanted to reduce this support, they could do so. If lower-level TCO operatives are responsible for rallying support among local populations, it could be fairly difficult for TCO leaders to effect change.

Because these assumptions are so uncertain, this portion of the IIP effort will have to be monitored and assessed very carefully to see whether course corrections are necessary or whether the entire effort needs to be scrapped.

Of course, the uncertainty in both sets of assumptions—the core assumptions about paths to popular rejection of the DLB and the assumptions about how TCOs might be made to reduce their support to the DLB—could be reduced with better understanding of the groups and individuals we want to reject DLB. **Target audience analysis can help you avoid (or diagnose) bad assumptions.** If we learn more about the TCOs and discover that, in the past, GOT has used the threat of increased law-enforcement attention to successfully pressure them into changing some aspect of their operations, then we would have more confidence in the plausibility of the related assumptions. By contrast, if we learn that the head of the Termina Triad is a cousin of the head of the DLB, and both come from a cultural background that stresses blood ties, we would be less sanguine about the prospects for the planned effort. Similarly, if we analyze the various groups in southern Takanwei and find that their grievances have little to do with a lack of security force presence and government services, that would weaken our confidence that increasing these things will motivate changes in behavior toward the DLB.

These two sets of assumptions serve as examples, but they are not exhaustive of the assumptions present in the logic model. The point of reviewing the connections between the different levels of the logic model is not to be exhaustive; it is to identify uncertain or vulnerable assumptions so that they can be watched. Similarly, identifying possible unintended consequences or disruptions to the logical flow makes us aware of these threats to success and allows us to watch for them and plan accordingly.

Again, some examples. Tasks 1, 2, 3, and 5 ("activities" in the logic model) all involve MISO efforts to use messages to convey information aimed at behavior change. Presumably, some of these messages will be transmitted as radio or television products. We could note the factors that could prevent such messages from reaching their intended audience. We should have identified some of the barriers or disruptors in our target audience analysis (e.g., which media outlets the audience turns to for information).

Other barriers or disruptors are more circumstantial, such as power outages due to flooding or DLB attacks on infrastructure, or a downed transmission or relay antennae, with the DLB seeking to control which stations are allowed to continue to broadcast. As another example, consider subordinate objective 1a2, "Populace rejects DLB to prevent access to new safe havens." When that outcome is achieved, what comes next? We hope that dejected DLB fighters will leave the area on their own, but we need to recognize that they might intensify their intimidation efforts instead, seizing needed support rather than receiving it from a willing populace. Other events that might cause efforts to go awry could include news of atrocities, either real or manufactured. If elements of MNF-DP committed atrocities, it would undermine and derail tasks (activities) 1, 4, and 5, at least. Furthermore, whether or not any MNF-DP forces commit atrocities, the DLB might accuse it of such crimes, backed by manufactured visual or physical evidence, staged atrocity sites, or falsified eye-witness accounts.

By identifying vulnerable assumptions, possible unintended consequences, and barriers to progressing from left to right across the logic model (whether due to circumstances or enemy action), you can identify factors that you might need to monitor as a possible warning that something is derailing the effort. In this way, it may be possible to develop contingency plans to surmount or contend with some of these barriers, if they are encountered.

*After too many hours working through the logic model and trying to spot things that could make it go awry or that could have unforeseen results, we in the IOWG were getting pretty tired of each other. There's a fine line between plausible unforeseen consequences or disruptions and stuff that is just off the wall. For example, when we were talking about how MISO broadcasts could be prevented from reaching their audiences, one of the guys started going off about solar flares! The rest of us groaned. That was just too much detail, and the likelihood is too low to include it. We already had a list of plausible reasons that a broadcast might not reach the intended audience: either people choose not to tune in or they can't tune in. Solar flares would fit in that latter category, along with likelier events, like power outages, jamming, or DLB capture or suborning of broadcast facilities so that the broadcast can't go out.*

*So, after some wrangling and a few unproductive arguments, we finally got around to a big list of possible spoilers and possible unanticipated consequences. We tried to rank them based on how concerned we thought we needed to be: How likely is this to happen, and how badly would it interfere with our efforts if it did happen? We'll use that prioritized list when we move on to planning data collection for evaluating and monitoring. We'll tackle that tomorrow. Right now, I need a break.*

## Be Thoughtful About What You Measure

The logic modeling process produced a host of inputs, outputs, outcomes, disruptors, and unintended consequences, all of which might be measured and monitored. A quick review of Table 4.3 reminds us that there are dozens of elements in the logic model. We further identified another fistful of possible disruptors or unintended outcomes—and both the logic model and the disruptors only cover objective 1! Collecting precise quantitative measurements for all of these elements would not be feasible for MNF-DP. Fortunately, the process of assembling and reviewing a logic model is helpful, because **logic models provide a framework for selecting and prioritizing measures**. Which of the things in the logic model really need to be measured? How well do they need to be measured? These two questions can be answered together by **considering the importance of candidate measures** and recognizing that we **only need to measure as precisely as required**.

Looking again at the final logic model for objective 1 (in Table 4.3), consider which things are important to measure and how precisely they need to be measured. We absolutely must measure all three levels of outcomes; we have to know whether our efforts are working. We need to measure our activities and their outputs; sometimes, those things can be measured jointly. For example, task (activity) 4 is "Reinforce messages with action," and the associated output is "arrests/capture in response to tips, improved provision of service." The activity is nebulous, so it does not require direct measurement. However, the output can be directly measured, and it should be. We also want to make sure that we are measuring things related to vulnerable assumptions to determine as quickly as possible whether those assumptions are holding up.

What level of fidelity is required in these measures? Although we need to measure all the objectives and many of the outputs and activities, **we do not need to measure all of them equally well**. In some cases, we really do want precise counts and measures. Fortunately, for some of those things, it is very simple to get precise data. For example, for the outputs of activities 3 and 4, pertaining to the tip line, we want to know exactly how many tips are called in each week (or month, or whatever we decide the reporting period should be), how many of those tips were validated by intelligence, and how many of those validated tips led to action (e.g., an arrest, a cache seizure). Fortunately, these things are all easily observed by the personnel operating the tip line and conducting operations related to the IIP effort, *provided they have been asked to keep track of them*. Beyond this, we do not need such precision. Sometimes, a yes/no answer is sufficient. That is often the case with activities, such as establishing a tip line (part of activity 3): *Yes*, we have activated the tip line; *no*, there has been a delay. Sometimes, yes/no is a good start, but if the answer is no, we want more detail. For example, to address intermediate objective 1a, "Deny DLB access to new areas/safe havens," it is just fine to record, "*Yes*, DLB has been kept out of new areas." However, "no, DLB is moving into new areas" requires more information: What new areas? At what rate? Of course, to be able to reach that initial *yes* with any confidence, someone will have to be tracking many different measures and indicators. In this particular example, that is the responsibility of intelligence; keeping track of the adversary is a big part of what the intelligence section does, so we do not need to worry about setting up the necessary collection and measurement efforts as part of IO assessments.

This does point to a general lesson, however: Sometimes, intelligence is already tracking the things we need to know. The intelligence staff is always heavily tasked and usually receives more requests for information than it can respond to. Requests that require new intelligence collection will compete for priority with such requests from all other staff sections. But requests for intelligence that is *already being collected* require less staff time and effort. Ideally, you will seek out existing intelligence collection activities to support your assessments.

So, how do we decide what to measure, and at what level of fidelity? Working through the different activities, outputs, and outcomes for objective 1 illustrates how to make this determination. We will work from right to left across Table 4.3, the logic model for objective 1. In the interest of efficiency, we will focus on elements of intermediate objective 1a (so, mainly the top half of the logic model).

The rightmost objective in the logic model ("Deny safe havens to DLB/isolate DLB from the populace") is the most important thing to measure, because whether or not you have achieved that objective is how you will gauge whether you have succeeded or failed. The specifications added when making an objective SMART will help you determine what and how to measure. Objective 1 has two components: denying safe havens to DLB and denying DLB support from the populace in potential safe havens. Both need to be measured. The first component is primarily spatial: Where are DLB camps, safe houses, and supply depots? Where are the group's leaders, fighters, and the other members based? Where are they free to move and operate? The spatial component of objective 1 is only of interest *over time*. We cannot assess whether the DLB has been denied safe havens now, at $t_2$, if we do not have at least one prior measurement (at some $t_1$) against which to compare. Our assessment will only be stronger if we measure at repeated intervals and **try to capture trends over time**.

The fully SMART version of the objective will explicitly state target thresholds, but we will know we have *failed* if the DLB retains a presence and freedom of movement in all safe havens in use at the onset of operations, and if it subsequently expands into new areas. We will know we have *succeeded* if the DLB does not move into new safe havens and if its existing safe havens are diminished by a targeted amount. A more complicated situation (and perhaps a form of mixed outcome) will occur if MNF-DP activities drive the DLB *out* of some safe havens and *into* new safe havens. Regarding measurement fidelity, measures must be sufficiently accurate and updated with sufficient frequency to be able to discern such changes. It would be pointless, for example, to report weekly updates on progress toward safe-haven denial if the underlying data were updated only monthly. Enemy presence and safe havens are very much within the realm (and part of the core competency) of the intelligence staff, so we should plan to rely on intelligence staff input to measure progress on the spatial component of objective 1. As discussed earlier, we will coordinate with the intelligence staff to learn whether it is already collecting information that meets our needs.

The second component of objective 1 has to do with denying DLB support from specified populations. The SMART version of the objective clarified "support" to mean tangible support—the provision of money, materiel, food, intelligence, and similar assistance. To sufficiently measure this component of the objective, we need to know, first, whether the DLB's support needs are being met and, second, which groups are providing that support. Information about adversary logistics is also something we can likely get from the intelligence staff, assembled from a range of sources, including detainee interrogations, overhead imagery, and human intelligence. We will coordinate with the intelligence staff to make sure it understands what we need and that what we need is already being collected. After-action reports from MNF-DP forces that engage DLB forces may help, too, to the extent that they report observed signs of shortages: limited ammo, poor nutrition, or equipment or clothing in poor repair. Though tricky to measure, intelligence will **try to triangulate data from different sources** to provide a more complete and accurate picture. By using multiple types of measurement or possible indicators, we will get a better sense of whether or not DLB tangible support is actually under pressure.

While we expect to get a summary measure from the intelligence staff of support available versus support required and where it is sourced, we will want to work with intelligence personnel to make sure they understand what we really need to know and why. Good understanding of not just whether the DLB continues to get support from various populations but also from which groups will help us discern progress toward the objective, refine our efforts, and focus more on the groups that are still providing support. Again, we want to coordinate all intelligence requests with J2 so that we minimize the extent to which we are asking for new or different collections. If we can find a way to get what we need from information that is already being collected, we are more likely to get it. And, if we limit our requests for new intelligence collection to a very few, very important things, we are more likely to get those, too!

Intermediate objectives 1a and 1b will be covered by the same kinds of reporting and measurement required for objective 1, provided that the data can be broken down between any new safe havens and existing safe havens. We need to be sure to make that distinction to guard against the undesirable possibility that our efforts amount to "squeezing the balloon"—pushing the DLB out of one area while prompting it to expand into another. If we were to see such a sequence playing out, we would want to redouble our efforts on tasks (activities) 1–5 in areas adjacent to prior safe havens, and we would want to focus checkpoints and patrols along possible lines of egress leading to potential new safe havens.

The subordinate objectives start to get toward things that will require additional thoughtful measurement. Subordinate objective 1a1 concerns physical interdiction, and the main measurements needed are covered by the spatial data for objective 1. If things are going well, areas that we are trying to prevent the DLB from entering will show no DLB presence. However, because we recognize that DLB infiltration is a possibility, we will want to have additional measures in place in case things do not go well and we need to make adjustments. These might be informal and low-fidelity measurements that we seek to collect only if things do go poorly. So, for example, if the intelligence staff reports new DLB presence in an area previously interdicted, we might seek to identify how those forces arrived there. This might involve closer examination of possible routes using overhead imagery, or information might come from detainee interrogations (if a DLB operative is captured in the interdicted zone, he or she could be asked how the group came to be there) or from after-action reports from patrols (e.g., Have patrols encountered DLB fighters, noticed subjects fleeing the patrols, or seen evidence of cross-country passage?).

Subordinate objective 1a2 concerns specific populations rejecting DLB presence. This is a contingent node in the logic model; that is, it becomes an objective to be measured only if certain other events take place. If the DLB does not try to move into an area, then groups in that area have no opportunity to reject the DLB. We could achieve intermediate objective 1a solely through subordinate objective 1a1 without this subordinate objective ever coming into play. If it does come into play, and if the DLB does try to move into new areas, we will want to measure and collect *both* indicators of success and indicators of failure, because we want to **avoid the temptation to collect data only on indicators of success**. These indicators are likely to come from interviews, feedback from key leader engagements, news reports, patrol after-action reports, or secondary reporting from non–MNF-DP organizations. Because it is a contingent node in the logic model, measurements will likely have to be taken after the fact. Too many areas might be possible new safe havens to arrange monitoring beforehand. If the DLB attempts to move into a new area—or if it succeeds in doing so—we can ask civil-military operations center partners, local leaders, and others what happened when the DLB came into town. Anecdotes from various incidents can be aggregated to create an overall summary. If no incidents of resistance (e.g., fights, demonstrations, shouting matches, threats, calls to tip lines) are reported, then various groups at least tacitly accepted the presence of the DLB and are providing support to the group. (Here, the presence of the DLB and a lack of incidents are indications of failure for this subordinate objective.) If, on the other hand, there are incidents, then it is likely that witnesses or participants will be willing to discuss what happened. Incidents alone are at least a partial indicator of success, but further data (and possibly further action) may be required. If the DLB is denied entry or support by a population, it may return in greater strength and force entry or coerce support. While full measurement of this subordinate objective may be both contingent and informal, measuring effectiveness in achieving it may require very close and timely monitoring. Knowing when the DLB has attempted or is attempting to move into a new area should trigger increased MNF-DP presence and support to relevant groups (activity 4) to reinforce and validate the various messaging efforts and to limit the DLB's ability to strong-arm the population.

*As we thought through measurement, we realized how important it was to know as soon as possible whether and when the DLB was moving into a new area. This was critical both for our informational efforts and for the physical efforts being integrated in pursuit of intermediate objective 1a. We worked with the J2 folks to make sure they had a good set of indicators and warnings for DLB moves into new areas. Turns out that this was something they were already watching, and they were happy to share their reporting with us. We built some new sequels on the branch plans in our IO plan to respond to those indicators, and then we elevated the issue to the full planning cell to make sure MNF-DP writ large would be ready to be nimble and responsive if the DLB tried to move into a new area. We certainly did not want to miss an opportunity to reinforce someone we had convinced to stand up to the DLB!*

Having considered measurement in relation to objectives, intermediate objectives, and subordinate objectives in our logic model, we turn now to measure activities and outputs. Here, we look at measurement for each of the five activity-output pairs supporting subordinate objective 1a2. The first two are activity 1, "Publicize MNF-DP and Takanwei PSF efforts to patrol, establish checkpoints, arrest, and combat DLB," and activity 2, "Encourage local populations in areas at risk to vocally and physically reject DLB incursions," which uses MISO to promote specific behaviors. The attendant outputs include the dissemination of the information and products and the receipt of the information by specific populations. Measuring activity 1 is easy. Did we disseminate press releases, give interviews to journalists, make MNF-DP broadcasts, and distribute MISO products? Of course we did, and the IRC staffs responsible for these efforts can measure them. They can even provide counts of each type of efforts. Their counts would be welcome, provide we do not end up thinking that more is necessarily better. Quantity does have a quality all its own, but once there is a sufficient level of dissemination (whatever that happens to be), that is enough. Most of the outputs are similar: How many of our products were disseminated? How many news sites, newspapers, or local television or radio shows used the press release or journalist interview content? Did those uses align with our objectives? Again, these are easy things to count, but someone has to be tasked with that counting and with deciding whether a radio or news item mention is favorable. The tricky output is whether or not specific populations received the information.

This is fundamentally a measure of exposure. Some efforts have built-in or simple measures of exposure. A website with a hit counter, for example, will tell you how many page views it has received. Fancy counters will track digital signature information and tell you each visitor's country or region (or, at least, where their Internet service providers are located). That does not get you an exact match to how many people in an audience of interest received a message, but it is not a bad indicator. Commercial television and radio companies' marketing departments will usually offer reach information for their broadcasts when they sell airtime, but these are often notoriously inflated. If elements of the desired information show up as threads in the social media exchanges of a certain group, you can be fairly confident that those individuals have been exposed to the information, but it is hard to confidently generalize from a few individuals on social media to a whole group. If you really want to know whether people have received some piece of information, you have to ask them. Surveys, interviews, or focus groups might be appropriate ways to ask members of a target audience whether they have received particular information. Thoughtful approaches to sampling (choosing whom to speak to) can make it so that you do not have to question very many people.

In some respects, this discussion comes back to **the importance of the candidate measure**. Is being highly confident that the target audience actually received your information worth the cost of a survey? If not, there are probably other alternatives to consider. Is there an existing survey that at least partially covers your audiences of interest? Are there other measurements that must or could be measured via a survey, and, if they are combined, does a survey become worth it? If so, consider adding a question or two to those surveys. Similarly, additions might be made to planned focus groups, or informal face-to-face engagements by patrolling security forces could be used to make informal assessments of exposure. Alternatively, assume that exposure is adequate, barring evidence to the contrary, and then try to measure possible interference. In looking for vulnerable assumptions, we identified several ways in which events or DLB actions might interfere with dissemination, including power outages, jamming, and seizing control of transmission stations. Did any of those things happen? If the intelligence staff knows you care about these things, it can be cued to report whether the DLB has jamming capability and whether there are any reports of its use, or whether enemy activity has included the seizure of any broadcast means. Information about power outages is likely available through the civil-military operations center, if you ask. Always balance the importance of a possible measure against its cost. And **do not conflate exposure with effectiveness**. Just because you establish with confidence that most of your target audience has indeed received the information you disseminated, that does not mean that it is going to respond to that information in the way you intended. You will have to measure that response, too. If things have gone well, the measurement of effectiveness will gauge progress at the outcome level—subordinate objective 1a2, specifically.

Activity 3 ("Encourage local populations in areas at risk to report DLB presence and activities to a tip line") also involves MISO promotion of a specific behavior, though in this case, the behavior is the use of a tip line. The dissemination and receipt measures will be similar to the previous two activities, with the added confirmation that information is being received if tips are being received. The portion of the output that concerns the establishment of the tip line is still trivial: Is the tip line established and staffed when it is supposed to be? Yes/no measurement is both easy and totally sufficient. Thinking about further measurement for the tip line reveals that part of the desired outcome has been left out. It should either have been specified as part of subordinate objective 1a2, or it should be its own outcome. For this to work, we want people to actually call the tip line and offer valid tips. Again, this is trivially easy to measure by those operating the tip line. How many calls were there? There should be a log, and calls should be easy to count. How many tips were validated by intelligence, and how many of those tips resulted in some kind of action? It is easy to develop a very clear sequence with easy-to-collect measures and compare against target levels, maybe even by region.

Activity 4 is "Reinforce messages with action," and the outputs call out two kinds of action: action on tips and improved provision of services. Action on tips follows the tip-line discussion from activity 3. While convincing target audiences to call a tip line is a nice start, that is not where the sequence ends. We do not want a "tip line to nowhere"; rather, we want the tips recorded, transmitted, validated, and (when appropriate) acted on. This serves the dual purpose of reducing DLB presence or capabilities (through action) and reinforcing the tip-line use (when tipsters see actions that might have been enabled by their tips). The same chain of measures easily collected as part of the top-line effort that ends with "number of validated tips acted on" is a good output measure for that part of the activity.

The activity for the CA/provision-of-services portion of activity 4 is easy to measure (acting MNF-DP elements need only report the things they do), as is the output. The output is "improved provision of services." Providing some sort of service that would not have otherwise been provided is probably a good thing, and it can be shown to be occurring based on measures of performance reported by the executing IRCs. However, there appears to be another missing outcome—perhaps another subordinate objective or an outcome between the output and the outcome. It is actually listed as part of the output: *improved* provision of services. Improved for whom, and relative to what? Should we be comparing services provided against zero, against the period right before we commenced operations, against our previous period of operations, or against a historical baseline? All have some potential merit but might have different implications for a progress report. What is desired should be part of the SMART review, which is why this point should have been specified as an outcome/objective. Services should have been specified: provision of electricity, sewage service, or drainage, for example. A baseline should have been measured, because **evaluating change requires a baseline**; then, a specific target level should have been set (with incremental progress expectations); and *then* progress toward that target could be measured. That measurement would require more (and more difficult) data collection. If we erected 22 power transformers and 16 km of high-tension electrical wires, that is an output and easily measured by those doing the construction. Change in the percentage of the population in an area with electrical service, on the other hand, is harder to measure. If accompanied by a modern power grid, with proper tracking for billing purposes, measurement would be as easy as asking the power company. If not, it might be estimated using power throughput draw combined with nighttime satellite data on illumination patterns. Because this aspect of the output/outcome has nothing to do with our IIP effort, we will not identify further possible measures.

Activity 5 is "Advertise CMO/security force successes, emphasizing success of cooperation and legitimacy of GOT and MNF-DP," so it will follow the same pattern for measurement as activities 1 and 2. What is missing, however, are some of the connections between the outputs and the outcomes. As noted during the logic model review earlier in this chapter, the vulnerable assumptions extend to the core logic of the logic model—namely, increasing security and services, plus increasing awareness of security and services, plus demonstrated effectiveness against the DLB in the area will encourage the specified populations to reject the DLB. How will we know if those assumptions are not holding? More importantly, how will we know where our effort is breaking down if those assumptions are not holding? We have planned to measure the outcome: the rejection of the DLB if it moves into a new area (as part of intermediate objective 1a). We plan to measure the increase in security and services as part of the activity and output measures. We plan to measure efforts to increase awareness as activity measures and success in increasing awareness as output measures. The gap is most likely between awareness of those good things among the specified populations and those populations actually engaging in DLB-rejecting behaviors (e.g., calling the tip line, creating incidents when/if the DLB tries to move into an area). Perhaps we need to extend the logic model. What other incremental steps might happen between awareness of services and security and the rejection of the DLB? What possible barriers did we identify in our logic model review? Can we measure any of those things?

As with the contingent node referenced earlier (subordinate objective 1a2), for some of the things we might measure, we might seek data only if there is a break in the chain of logic. If this does not occur (that is, if everything works and follows the logic model), we will not try to collect those measures. However, if we see positive measures of performance and positive outputs for all activities, but we do *not* see the desired rejection behaviors, we will know that there is a break in our assumptions and we need more information. Under those circumstances, we might try a range of measures that do not require great investment to collect to try to triangulate what might be wrong. Then, anything we identified as possible barriers during our logic model review can be turned into hypotheses and considered. For example, we might hypothesize that DLB intimidation is preventing selected populations from effectively rejecting the group. What might we see if that were the case? We start by considering things we might already be collecting: Where, at the objective 1 level, is the DLB getting its support? If support is still flowing from the population but indicators suggest it is now coerced rather than given freely, we have evidence to support the hypothesis and guidance for what we might do about it. Are DLB-caused civilian casualties up? That could also be an indicator of increased intimidation. If existing data do not help, consider short-term emergency measurements. Why aren't even intimidated individuals using the tip line? Maybe the local population believes that the DLB has infiltrated the tip line program and that people who provide tips will be targeted for reprisals. Talk to interpreters, human intelligence sources, and civilian key leaders and ask them why tips have dried up. If these sources confirm this hypothesis, then you can start to work on solutions. When there is a break in the chain of logic, you do not have to completely and confidently diagnose it and work toward a solution. Conceive a number of different possible explanations, measure what you can, however you can, and if the evidence even partially supports one of your candidate alternatives, move out to try to fix that problem concurrently with better data collection to confirm the problem.

Where there are vulnerable assumptions that are hard to break out, or when it is resource-intensive to measure the subcomponents, remember to *fail fast*. That is, monitor the links in the chain of logic adjacent to the vulnerable assumptions to see whether they are moving together or not. If they seem to be working, that is well and good. However, if they seem disconnected, be ready to take aggressive short-term measurements to figure out what is causing the disconnect and to try to fix it.

# CHAPTER FIVE

# Conclusions and Review

*It was hard work and led to some long hours and some shouting matches in the IOWG, but I'm glad we did it. I now have confidence that we are going to meet our objectives, that we'll be able to report progress, and, most importantly, that we'll be able to adjust and refine efforts to meet changing circumstances (and changing assumptions) as the mission requires. I'd rather do the hard work at the front end and plan for strong assessment reporting than have something go wrong midstream and not be able to figure out how to fix it. I've still got to convince the planning lead that some of the things we need measured should be Commander's Critical Information Requirements so that we can get J2 to focus on them, but I've got compelling arguments. The DLB is in trouble; we're going to put the hurt on them. Fnorky, out.*

This report used an artificial context and realistic but fictional mission to provide a worked example of planning for assessment and monitoring for JTF IIP efforts as part of an overall IO campaign. The example demonstrated the application of core assessment principles and walked through some of the challenges and difficulties that practitioners will face as part of that process. The example further demonstrated some strategies to employ to surmount those challenges and difficulties.

Specifically, readers (and practitioners) are reminded that **effective assessment starts in the planning phase**, and, above all else, **assessment must support decisionmaking**. Furthermore, **effective assessment requires clear, realistic, and measurable goals**—goals that are SMART (specific, measurable, achievable, relevant, and time-bound) and specify the observable behaviors sought and at what target threshold. These goals or objectives need to at least imply what failure would look like, and they should be able to be broken into smaller subordinate objectives or sequential steps to make assessment easier. Evaluating progress against these objectives requires some kind of baseline measurement. After all, **effective assessment requires a theory of change or logic of the effort connecting activities to objectives**, including planned inputs, activities, outputs, and outcomes. Logic models can help identify possible constraints, barriers, or unintended consequences to planned activities, and logic models can either start small and grow or start big and be pruned. Good target audience analysis can help avoid bad assumptions in logic models, and a "fail fast" implementation approach can also reveal flawed assumptions and help you correct them. Finally, **be thoughtful about what you measure**. Logic models can provide a framework for selecting and prioritizing measures. When choosing measures, consider the relative importance of different candidates, and be sure to collect indicators of both success and failure. Measures should not conflate exposure and effectiveness where messaging is concerned, and they should try to capture trends over time. Use multiple data sources to triangulate information where possible, but conserve scarce resources by only measuring as precisely as required.

The worked example presented a detailed walkthrough of part of the assessment planning process. Throughout this process, those planning assessments for IIP efforts should follow these steps:

1. Review and refine objectives (including intermediate objectives or subordinate objectives) to ensure that they satisfy the SMART criteria.
2. Build (and then refine) a logic model or other articulation of the logic of the effort supporting the objectives.
3. Plan data collection and measurement to support assessment, using the logic model and the list of vulnerabilities to help prioritize measures.

Consider building your logic model backward, listing objectives from top to bottom and right to left. Then, build in identified tasks and activities on the left, add additional activities and identify the outputs of those activities, and identify possible constraints, barriers, disruptors, or unintended consequences that might keep the effort from successfully following the logic model. Finish by reviewing all the connections and make sure they work logically, revising them until they do.

# Telecommunications and Media in Takanwei

## Telecommunications and Information Networks

The broadband market in Takanwei is one of the more robust networks in the region, with significant penetration throughout the country. Takanwei's telecom industry is state-owned, as is this case in many other countries in the region. Nearly all of Takanwei's telecom services fall under the government-controlled Dunaracom monopoly: fixed and mobile telephone service, Internet access, wifi, prepaid data network cards, corporate telecom services, and satellite service. Because all services must be approved by the state, the system is rife with inefficiencies and corruption, with months-long waits for service in certain regions of the country (mostly in the remote interior). The state has shown a typical reluctance to privatize or submit to competition, for fear of losing telecom revenue streams and being at the mercy of foreign investors for pricing and maintenance. Even with a state monopoly, investment opportunities are promising, especially in new technologies in the wireless field. Takanwei, literally cut in half by its dominant central mountain range, could benefit from future advances in wireless communication, especially over the more costly option of fiber-optic cable. It is estimated that wireless penetration in Takanwei has reached more than 87 percent, and the government has ambitious plans to ensure that the entire country has wireless network access by 2024.

## Newsprint, Radio, and Television

Takanwei has two daily newspapers, one based in the capital (Port Talbuk) and the other based in the southern city of Mezi, the country's second-largest urban center. Although both papers are government-owned and -operated, they reprint much of their content from world news sources. Local and national news is censored, and newspaper circulation is generally limited to government offices, embassies, education centers, and the country's elites. With a literacy rate of 55 percent, most people receive their news from television (two stations) and radio (six stations). All these outlets are state-owned and -operated. Radio is the primary means to hear news or share information, and several stations have call-in hours and typical soap opera–style shows that are very popular with the population. Those who can afford satellite dishes are able to receive world news instantaneously and uncensored.

# Timeline and Road to Crisis, January–July 2022

### January Key Events
- DLB activity in Arpanda focuses on planning unknown offensive operations.
- DLB activities in Takanwei are concentrated primarily in smaller communities in the south of the country and include clashes with the Takanwei PSF. The DLB attacks a police outpost in Port Cook, Takanwei. The attack requires deliberate planning and detailed analysis of the PSF outpost's daily patterns.
- DLB maintains numerous camps in Takanwei. The discovery of maps of PSF training facilities and lists of training locations and times indicates that the DLB has deliberately targeted the PSF.
- There is widespread flooding in the southeastern portion of Takanwei.

### February Key Events
- Arpandian security forces and the U.S. Coast Guard capture a vessel carrying arms and explosives that is approaching the eastern end of the straits.
- The DLB's leadership announces a new maritime strategy to interdict merchant shipping bound for the Straits of Arpanda as far as 150 nautical miles from the straits.
- The Arpandian and Takanweian foreign ministers meet to discuss bilateral strategies to counter the DLB threat.
- DLB activity in Arpanda focuses on interrupting traffic in the straits.
- The DLB obtains suitable vessels to interdict merchant shipping bound for the Straits of Arpanda: a refueling vessel that has disappeared from the port in Bonaire, Arpanda, and an ocean tug pirated from Landakan, East Polinae.
- Takanweian citizens report DLB activity along the Mimi River.
- DLB propaganda efforts in Arpanda focus primarily on threatening traffic in the straits and disrupting shipping while courting popular support for the DLB.

### March Key Events
- DLB activities include IED and maritime vehicle–borne IED construction, Q-ship arming, murder, and kidnapping.

- Takanweian groups report that DLB propagandists are recruiting and voicing an anti-capitalist agenda.
- A Takanweian patrol discovers bodies in Port Cook; reports indicate that the victims refused to join the DLB.
- Takanweian citizens report DLB activity along the Mimi River.
- Local police in Blackshores find remnants of IED materials in a vacant boathouse, including explosives, cell phones, wires, and tape.

## April Key Events

- DLB activity focuses on disrupting shipping in the Straits of Arpanda, including GPS and VHF jamming of commercial vessels in and near the straits.
- The DLB launches a new social media campaign seeking to gain 1 million supporters for its cause.
- Suspicious personnel conduct reconnaissance missions in the vicinity of the Darguli Narrows of the Straits of Arpanda.
- Takanwei PSF capture a semisubmersible in the Dunarian Sea containing DLB documents and explosives.
- The DLB attacks the main police station in Port Windalay, Arpanda.
- An attack is reported on a merchant vessel en route to the Straits of Arpanda, but the method of attack cannot be confirmed. Reports suggest an explosion, possibly from a mine or an IED planted on another vessel.

## May Key Events

- A merchant vessel in the Straits of Arpanda is fired upon by a small, unknown vessel. An explosion is reported aboard a Dutch cargo ship in international waters that is en route to the Straits of Arpanda.
- Another merchant vessel is fired upon in the Straits of Arpanda and outruns the unknown aggressor.
- Shipping companies consider canceling scheduled transits of the Straits of Arpanda.
- The King of Arpanda directs internal security Task Force Yi activation.
- The DLB engages in small-arms and rocket-propelled grenade attacks from both sides of the Darguli Narrows; no serious damage to ships is reported.
- An explosion is reported at an Arpanda City Metro Red Line station; there are no injuries, but the Metro remains closed for more than 48 hours.
- An Arpandian patrol boat is struck by what is believed to be a waterborne IED. Four crewmembers are severely injured; the patrol boat is towed to port but remains out of service indefinitely.
- The Straits of Arpanda Transit Authority's English website experiences several homepage defacements.
- Monsoons cause widespread flooding in Takanwei and Arpanda.

**June Key Events**

- The DLB claims responsibility for attacks in Arpanda and in and near the straits that destroy extensive maritime mining capabilities and threaten the straits. The group demands a $10 million ransom to discontinue its attacks on straits shipping and industry.
- The King of Arpanda sends a letter to regional leaders and the President of the United States advising them of the threat to the Straits of Arpanda.
- The governments of Takanwei and Arpanda inform the UN Security Council of the threat the DLB poses to the region and the straits.
- The Chairman of the Joint Chiefs of Staff issues a warning order to U.S. Eastern Command.
- The DLB claims responsibility for an explosion aboard a Greek tanker en route to the Straits of Arpanda.
- Eleven nations advise the UN Security Council that they will contribute forces to MNF-DP if the council passes the resolution.
- The UN Security Council passes Resolution 15080 authorizing the U.S.-led MNF-DP to assist Arpanda and Takanwei and to protect the straits.
- The DLB launches a radio propaganda campaign in Mezi.

**July Key Events**

- DLB activity in Arpanda focuses on planning unknown offensive operations.
- DLB activities in Takanwei focus primarily on the Mimi River and Blackshores and include clashes with the Takanwei PSF. The DLB attacks a police outpost in Port Cook, Takanwei. The attack requires deliberate planning and detailed analysis of the PSF outpost's daily patterns.
- The DLB maintains numerous camps in Takanwei. The discovery of maps of PSF training facilities and lists of training locations and times indicates that the DLB has deliberately targeted the PSF.
- A denial-of-service attack disrupts GOKA websites.
- The U.S.-led MNF-DP states that its objective is to eliminate the DLB threat to the Straits of Arpanda.

# References

Hubbard, Douglas W., *How to Measure Anything: Finding the Value of "Intangibles" in Business*, Hoboken, N.J.: John Wiley and Sons, 2010.

Mertens, Donna M., and Amy T. Wilson, *Program Evaluation Theory and Practice: A Comprehensive Guide*, New York: Guilford Press, 2012.

Osburg, Jan, Christopher Paul, Lisa Saum-Manning, Dan Madden, and Leslie Adrienne Payne, *Assessing Locally Focused Stability Operations*, Santa Monica, Calif.: RAND Corporation, RR-387-A, 2014. As of February 7, 2017:
http://www.rand.org/pubs/research_reports/RR387.html

Paul, Christopher, and William Marcellino, *Dominating Duffer's Domain: Lessons for the 21st-Century U.S. Marine Corps Information Operations Practitioner*, Santa Monica, Calif.: RAND Corporation, RR-1166-1-OSD, 2016.

———, *Dominating Duffer's Domain: Lessons for the 21st-Century U.S. Army Information Operations Practitioner*, Santa Monica, Calif.: RAND Corporation, RR-1166/1-A, 2017.

Paul, Christopher, and Miriam Matthews, *The Russian "Firehose of Falsehood" Propaganda Model: Why It Might Work and Options to Counter It*, Santa Monica, Calif.: RAND Corporation, PE-198-OSD, 2016. As of February 7, 2017:
http://www.rand.org/pubs/perspectives/PE198.html

Paul, Christopher, Jessica Yeats, Colin P. Clarke, and Miriam Matthews, *Assessing and Evaluating Department of Defense Efforts to Inform, Influence, and Persuade: Desk Reference*, Santa Monica, Calif.: RAND Corporation, RR-809/1-OSD, 2014. As of February 7, 2017:
http://www.rand.org/pubs/research_reports/RR809z1.html

Paul, Christopher, Jessica Yeats, Colin P. Clarke, Miriam Matthews, and Lauren Skrabala, *Assessing and Evaluating Department of Defense Efforts to Inform, Influence, and Persuade: Handbook for Practitioners*, Santa Monica, Calif.: RAND Corporation, RR-809/2-OSD, 2014. As of February 7, 2017:
http://www.rand.org/pubs/research_reports/RR809z2.html

U.S. Joint Chiefs of Staff, *Joint Operation Planning*, Joint Publication 5-0, Washington, D.C., August 11, 2011.